THE WHEELING SUSPENSION BRIDGE

A Pictorial Heritage

THE WHEELING SUSPENSION BRIDGE

A Pictorial Heritage

By Emory L. Kemp & Beverly B. Fluty

sponsored by

Pictorial Histories Publishing Co.
Charleston, West Virginia

LIBRARY OF CONGRESS
CATALOG CARD NUMBER 99-75116

ISBN 1-57510-066-5

First Printing: October 1999

Layout: Stan Cohen
Cover Graphics: Mike Egeler, Egeler Design

Printed in Canada

DEDICATION
FOR JANET AND BILL

Pictorial Histories Publishing Co., Inc.
1416 Quarrier Street, Charleston, West Virginia 25301
(304) 342-1416

Acknowledgments

The Wheeling Suspension Bridge has been of compelling interest for the authors for more than three decades. During this time we have been assiduously collecting materials on the history and engineering of the bridge. Along the way we have been indebted to many who have unearthed information on the bridge and its builder, Charles Ellet, Jr. Our research has ranged from Britain, France and in the United States at the University of Michigan, West Virginia University, the Smithsonian Institution, Rensselaer Polytechnic Institute, and local collections in Wheeling.

For this book on the occasion of the 150th anniversary of the bridge, we are indebted to our special friends at Oglebay Institute Mansion Museum, Holly McCluskey, Director; John Artzberger, Senior Curator; Sandra Porter, Reception, and an indispensable helper, who is the Administrative Assistant, Travis Zeik. We are most grateful to American Bridge for their generous support with this project. We also thank many others—Alan Behr, Elizabeth Bell, Jackie Bence, Michaelle Blum, U.S. Senator Robert C. Byrd, Robert De Francis, Charles Flynn, Hydie Friend, Cecil W. Gwinn, Wendy F. Hinerman (nee Fluty), Louis Horacek, Joellyn B. Kemp, Mark A. Kemp-Rye, Eric Merrill, Elizabeth B. Monroe, Heather Moore, Peter Samuel, Gerald Reilly, Mina Roberts, Donald Sayenga, David Simmons, Larry Sypolt, Robert Vogel, Eugene "Chip" West, Richard West, Joan E. Woodfin, William Worthington, and Gary Zearott. We are particulary pleased to have the book join a highly successful series published by Stan Cohen. Without his help and encouragement, the timely publication of this book would have been impossible.

E.L.K.
B.B.F.

Foreword

During the depths of the Great Depression, President Franklin Delano Roosevelt arrived by train in Grafton, W.Va., to view the Tygart Dam, one of the first large-scale public works under the New Deal. The president was criticized for his New Deal projects and was expected to defend his programs. In ringing terms he told the crowd that he had come not to defend the New Deal, but to proclaim it!

In recognition of the Wheeling Suspension Bridge's 150th Anniversary, we have come to proclaim it as one of America's great antebellum structures. With its completion, American engineers lead the world in long span suspension bridges for more than a century. Its designer, Charles Ellet Jr., introduced to America the French system of wire cable suspension bridges with his 1842 Fairmount Suspension Bridge. He promoted vigorously the idea of this new structural type for more than two decades. His masterpiece and the culmination of his promotional efforts was the erection of the Wheeling Suspension Bridge, completed in 1849.

Through photographs and drawings we hope to convey a sense of the role played by the bridge during its 150-year life, not only as a great engineering icon, but how it forged links in transportation and commerce in Wheeling. The text and illustrations are intended to present a context for the bridge in terms of the history of suspension bridges, but equally important, the influence it has had on the lives of the inhabitants of Wheeling and the upper Ohio Valley over the years. Few great 19th century public works have secured the public affection and respect as the Wheeling Suspension Bridge.

E.L.K.
B.B.F.

Table Of Contents

The Prelude: An Ancient Technology Transformed 1

Charles Ellet, Jr. 7

Building the National Road 13

The Wheeling Suspension Bridge 19

1850-1860: Court Case and the 1854 Windstorm 29

Civil War and West Virginia Statehood 41

Wheeling: 1870-1900 45

Wheeling and Its Bridge During the 20th Century 53

Wheeling Bridge Chronology 76

Selected Bibliography 78

About the Authors 79

The Prelude: An Ancient Technology Transformed

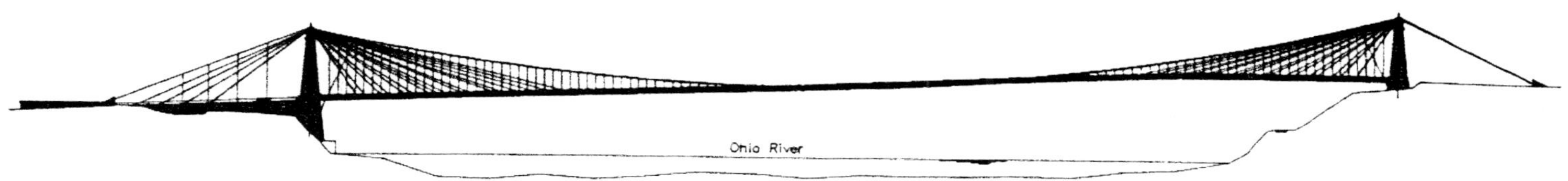

On October 20th—one hundred and fifty years ago—the Wheeling Suspension bridge was completed to great public acclaim. At more than 1,000 feet, it was the world's longest clear-span bridge. It was also a powerful symbol for the people of Wheeling, representing the long cherished dream of being a transportation hub on the upper Ohio Valley. With the bridge, the upper limit of navigation in the summer on the Ohio River, and the later arrival of the Baltimore and Ohio Railroad, Wheeling's population looked forward to a bright future where this city would be the premier port on the entire upper Ohio River. The vision was shared by others beyond the population in the northern panhandle. It was reinforced by the U. S. Treasury Department's decision to build a Custom House, which would serve not only for the U. S. Custom Service, but also for the U. S. Postal Service and the Federal Judiciary. This building was begun in 1854 and completed in 1859. And it was in this building, in the Court House Room, that the state of West Virginia was born on June 20, 1863.

The designer of the suspension bridge was Charles Ellet, Jr. (1810-1862). After securing the contract for the bridge, the Wheeling and Belmont Bridge Company provided a subvention for the publishing of his report. It was in this report that a drawing for the bridge, as Ellet envisioned it, was produced, showing an elevation, plan, and interesting details of the bridge. Prominently displayed in this drawing is a representation of a bridge in Latin America, which was published by Alexander von Humboldt. The illustration represents Ellet's retrospective on the origins of the suspension bridge, and we too, must turn our attention to the predecessors of the Wheeling Bridge to provide an adequate historical context. After all, the bridge did not arise like Minerva from the head of Zeus.

The Wheeling Suspension Bridge represents the transformation of a very ancient technology into a powerful new medium to serve the 19th century's Industrial Revolution in America. Suspension bridges arose in two very similar but isolated alpine areas of the world: the first, the Himalayas, and the second, the great chain of Andean Mountains in Latin America. In both these locations, suspension bridges were constructed of organic material such as vines, ropes, and even leather. They provided passage not only of pedestrian traffic but also of pack animals. In the case of the Andean suspension bridges, much information was transferred back to Europe by the Spanish conquerors. As early as 1535, we have information available in Europe on various suspension devices used by the Incas in their elaborate road system. Alexander von Humboldt was only one of a number of explorers in Latin America who commented on and illustrated suspension bridges found there.

The Europeans learned of the Chinese development of suspension bridges on their side of the Himalayas, and similar information was received on Indian suspension bridges, particularly in connection with the development of Buddhist trails connecting China with India. Very early on, the Chinese substituted iron for organic cables. These chain link bridges represent one of the first applications of iron anywhere in the world. From 65 A.D. to the beginning of the 18th century, there are records of 24 such iron bridges erected in the Himalayas. Information on these bridges was transferred to Europe by a number of explorers, but principally by Jesuits such as Athanasius Kircher.

Temporary military bridges were featured in a number of military campaigns in Europe several centuries ago, the most famous of which was the bridge at Alcantara in Spain, erected as part of Wellington's great Peninsula campaign and spanning an arch on the Roman bridge that had been destroyed by French forces.

Thus, in Europe, there was information on this structural form available from the 16th century. At the end of the 17th century, French mathematicians had solved the riddle of the suspended string and produced an equation for the catenary curve. The mathematical problem

The bridge at Penipe across the Chambo River in Latin America represents a legion of primitive suspension bridges erected in the Andes. Because the cables were made of vegetable matter fashioned into ropes, they had to be renewed frequently. Ellet featured this illustration in his report of 1847 on the Wheeling Bridge to remind readers of the ancient origin of these structures. This illustration came from Alexander von Humboldt's *Research Concerning the Institutions and Monuments of the Ancient Inhabitants of America*, London, 1814.

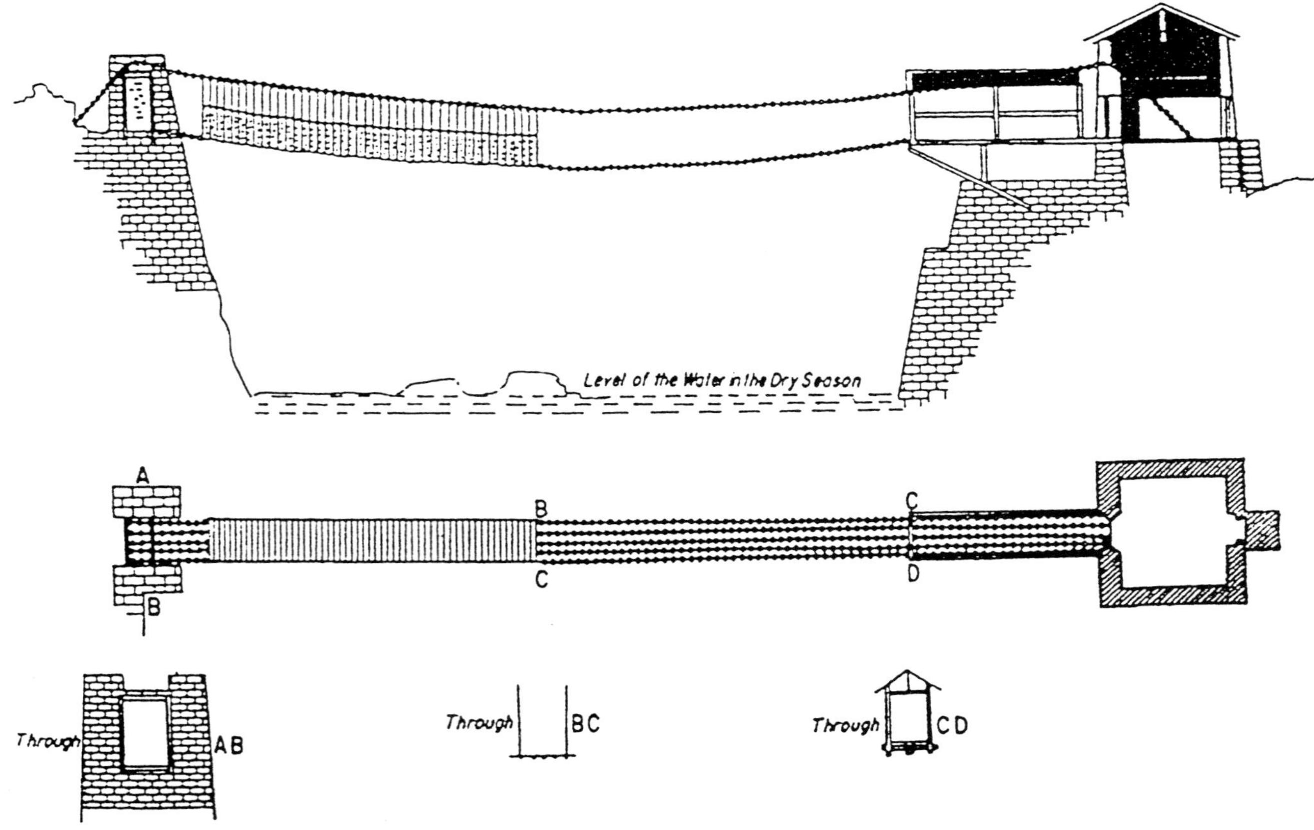

The peoples of the Himalayas produced not only fiber cable suspension bridges from ancient times but were the first to use iron chains. The Chuka Bridge over the Tehintchieu River in Bhutan featured iron chains with the deck laid directly on the cantenary chains. This bridge was reported by Samuel Turner in *An Account of an Embassy to the Court of the Teshoo Lama in Tibet*, London, G&W Nicol, 1800.

of the suspended string with individual loads resulting in a funicular polygon had also been solved. In the hands of the French, this theoretical information was transformed into the arch, which is the inverse of the suspended string. It provided engineers, for the first time, a theoretical basis for proportioning masonry arch structures. In addition to the theory of the suspended string and the funicular polygon, a considerable amount of work was done in the 18th century in terms of the strength of materials. Such information was provided in an encyclopedia produced in Philadelphia in the 18th century. Thus, all the elements were available shortly after the formation of the American Republic, which would provide the necessary background for the design and erection of suspension bridges.

With well-established practices of engineering in Europe and Britain, one would surmise that the modern suspension bridge had it origin across the Atlantic. That, however, was not the case. Judge James

A typical Finley chain bridge at Lehigh Gap, Pennsylvania, over the Lehigh River; completed in 1826. The timber-framed towers are covered in similar manner to timber covered bridges to protect the main timber frames from the weather. KEMP COLLECTION

Lehigh Gap Bridge, 1826, showing the link chains and wrought iron suspenders supporting a timber deck. PALMERTON, PA. CAMERA CLUB

Finley is credited with developing the first modern suspension bridge, in an undeveloped area of western Pennsylvania, settled first by Ulstermen from Northern Ireland. He was a near neighbor of Albert Gallatin, who built his home at Friendship Hill overlooking the Monongahela River. Finley also served in the Pennsylvania Assembly and was aware of developments on the East Coast. It was a time when

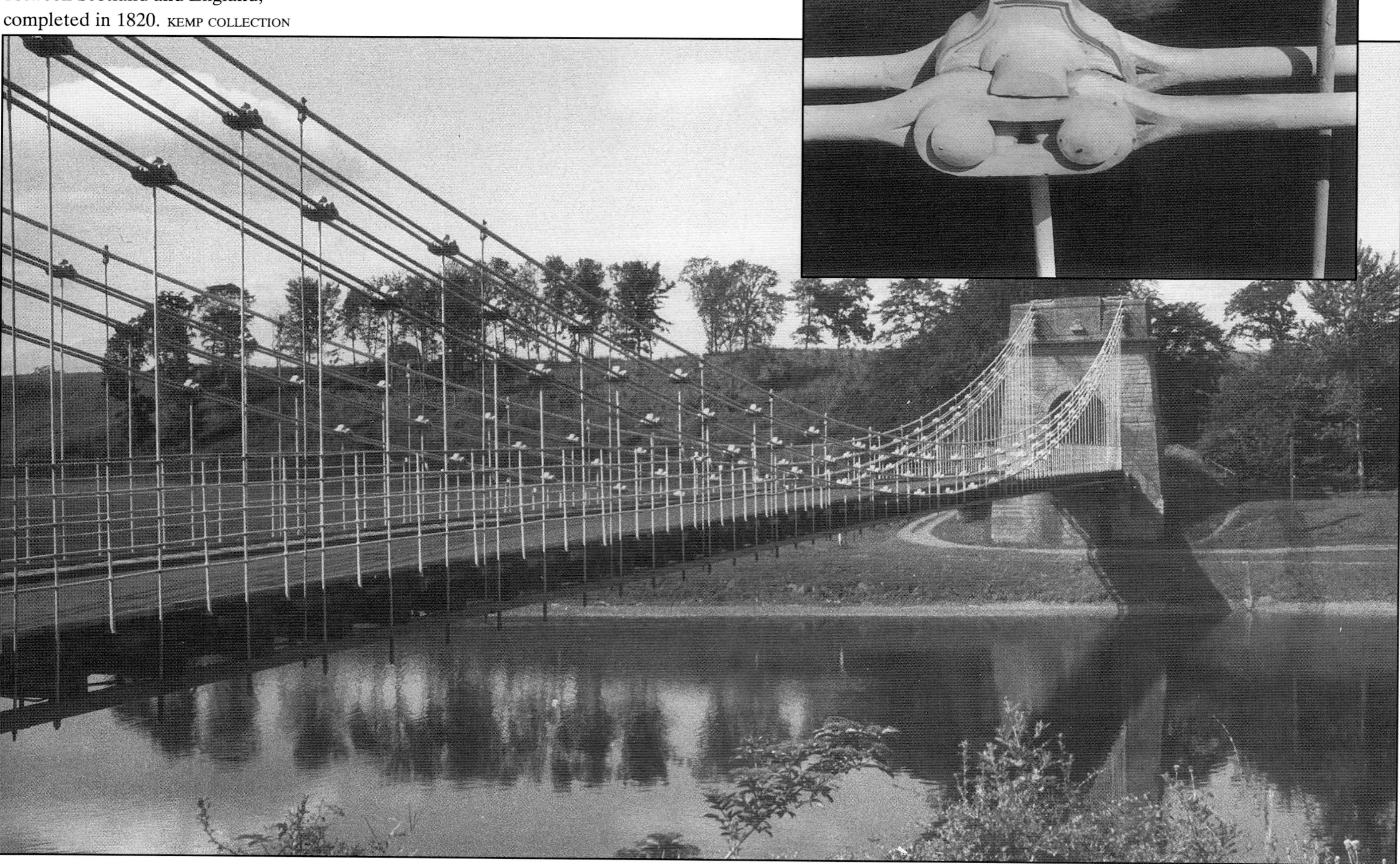

Samuel Brown's elegant Union Bridge over the Tweed River between Scotland and England; completed in 1820. KEMP COLLECTION

Detail of the chains of the Union Bridge. The coupling links support a vertical suspender as well as pinning the bars of the main chain together. KEMP COLLECTION

the Internal Improvements Movement, just beginning, sought to develop the rich natural resources beyond the Appalachian Mountains, and ushered in a time of intense competition between the various East Coast ports stretching from Boston to Norfolk, Virginia. One of the early proponents of this movement was none other than Thomas Paine, who was promoting iron bridges shortly after the American Revolution. It may be that Finley, too, was taken up with the idea of the Internal Improvement Movement, which would certainly involve the development of a road system across the nation. Gallatin felt the Internal Improvements Movement would be the means of developing a more secure Union amongst the disparate 13 colonies, which were being forged into a new nation.

In any case, the first modern suspension bridge was constructed according to a design of James Finley across Jacob's Creek on the turnpike from Uniontown to Greensburg, Pennsylvania. This diminutive bridge had a span of 70 feet and yet, represented all of the elements of the modern suspension bridge: iron chains for the main supporting elements, towers to support the chains, anchorages to secure the back stays of the bridge, a stiffening truss, and level deck suitable for wheeled vehicles. Later, Finley wrote a paper in *Portfolio Magazine,* at about the same time that Gallatin published his famous work on internal improvements, in which Finley discusses each of these elements of the bridge and presents a remarkable experimental method for determining not only the geometry of the main chains, but also the force in the chain from the center of the bridge to the towers. Examination of this work confirms that his method did indeed provide very useful and accurate information for the use of local designers. Finley became what today would be called a consulting engineer and was responsible for the design of numerous bridges from 1801 until his death in 1828.

His colleague, John Templeman, in a patent of 1810, suggested that more efficient main chains could be constructed of eye bars, rather like a large bicycle chain, or by using wire built up in parallel strands. In 1816, a pedestrian bridge was built across the Schuylkill River by the firm of Hazard and White, who were in various businesses including the production of wire for the textile industry. At the same time in Galashiels, Scotland, Richard Lees built a similar bridge to provide access for textile workers over the Galawater River

Information on the Finley development of suspension bridge design and construction was available in Britain, and early on taken up by Sir Samuel Brown and the famous civil engineer, Thomas Telford. The British succeeded in building a number of spectacular early suspension bridges, including Brown's Union Bridge of 1820, Telford's suspension bridge over the Menai Straits in Wales, completed in 1826, and William Tierney Clark's suspension bridge of 1831 over the Thames at Marlow. Under the leadership of Samuel Brown, the British preferred to use coupled bars or rods for a number of small-span bridges, but later favored flat eye bars that made up massive chains for long span suspension bridges. This became the symbol of British bridge development.

During this period the French became intrigued with British developments and sent the famous mathematician and engineer Claude L. M. Navier to Britain to examine current work, which included a visit to the Menai Bridge. He returned and provided a report, a virtual textbook of suspension bridge technology. At about the same time, Marc Seguin, who was the nephew of Montgolfier of hot air balloon fame, erected, in 1823, the first wire suspension bridge in France, also to provide pedestrian traffic for his workers at Annonay. This was the prelude to Ellet's involvement in suspension bridges and represents the state-of-the-art at the end of the 1820s.

Thomas Telford's Menai Suspension Bridge with a span of 580 feet, completed in 1826. The illustration is from an early French engraving. KEMP COLLECTION

Charles Ellet Jr.

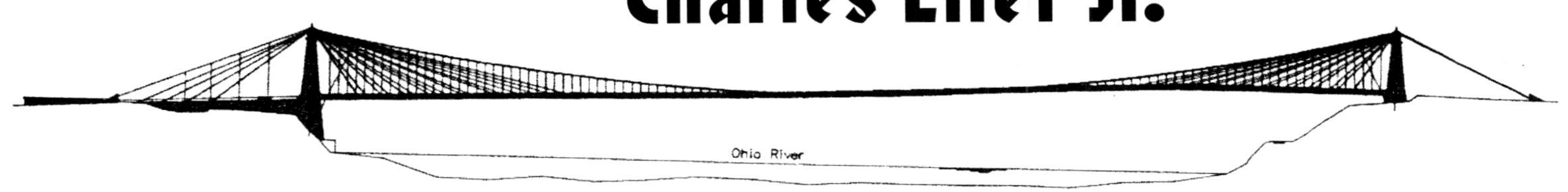

U.S. ARMY CORPS OF ENGINEERS, HUNTINGTON DISTRICT ARCHIVES

Charles Ellet, Jr. was born at Penns Manor in Philadelphia in 1810. Rather than remain on the farm, which was his father's wish, the young Ellet left the family home and joined a survey on the North Branch of the Susquehanna River in 1827. A year later, he was employed on the Chesapeake and Ohio Canal under the legendary Benjamin Wright, who is credited with being the father of civil engineering in the United States. He rose rapidly, in view of his age and lack of formal education, to the position of assistant engineer in 1829. Despite the long hours and arduous work on the canal, Ellet spent his evenings and other free moments learning French and studying mathematics. In 1830, he resigned his job with the C&O Canal, and although invited by his family to move to Illinois where prospects for an engineer appeared extremely good, traveled to France in June, armed with letters of introduction from the U.S. Ambassador and none other than the Marquis de Lafayette, and began attending lectures at the world famous École des Ponts et Chaussées. At the École, he had the opportunity of attending lectures by Navier and meeting leading engineers of road and bridge construction in France. He attended two sessions, one in the fall, and the other in the spring of 1831. Following these intense studies, he undertook a tour of southern France and Switzerland.

Upon his return to the United States in 1832, Ellet became a proponent of the French wire suspension bridge. Relegating earlier work by his own compatriot, James Finley and others, to the category of primitive architecture, he sought over many years to promote the design and construction of long-span wire bridges. Shortly after his return, he transmitted a design for a suspension bridge over the Potomac at Washington, but it arrived too late to be considered by the committee. He moved on to become the Assistant Engineer to the Western Division of the Utica and Schenectady Railroad in 1833, and by 1834, had become a surveyor for the New York and Erie Railroad in New York State. In 1836, he moved to the James River and Kanawha Canal, where he was appointed Chief Engineer. He remained, through these moves, a ceaseless promoter of long-span suspension bridges.

Never one to suffer fools gladly, and having a rather abrasive character when it came to his professional dealings, it is not surprising that he was relieved of his position of Chief Engineer on the James River and Kanawha Canal in 1839. This gave him the opportunity to go to the west where he spent some time with his brothers Alfred and Edward at Bunker Hill, Illinois. It was during this time that he prepared an elaborate design for a bridge across the Mississippi River at St. Louis. It was one of the most extensive reports of his career and featured a 3,000-foot, three-span suspension bridge across the mighty Mississippi. It would not be until well after the Civil War that the well-known Eads Bridge became the first crossing of the Mississippi at St. Louis.

He continued to promote suspension bridges along the Ohio Valley during this period, and in 1841, he won the contract, after much negotiation and some doubtful maneuvering on the part of parties involved, for the Fairmount Bridge across the Schuylkill River at Philadelphia. The famous Colossus wooden bridge there burned in 1838, and Ellet's bridge, using the abutments of the original covered bridge, replaced this notable structure. With the completion of the bridge in the spring of 1842, Ellet had succeeded in designing—in every detail—a French suspension bridge on American soil. It catapulted him into a leading position with regard to suspension bridge building.

It was during the 1830s that Ellet first became interested in the possibility of building a long-span wire suspension bridge across the Ohio at Wheeling, as a result of the reorganization of the Wheeling and Belmont Bridge Company in 1836. As early as 1846, he was in stiff competition with John A. Roebling for the construction of a dual rail and highway bridge across the Niagara Gorge. Persistence had paid off when he was awarded the Niagara suspension bridge contract in 1847. Earlier that year he was appointed to build the Wheeling suspension bridge.

All did not go well at Niagara. On advice of his lawyers, Ellet, who was something of a showman, charged tolls to ride across the gorge in a four-person iron basket or on his temporary bridge. He retained the tolls, to the annoyance of the company, which led to his final dismissal from the project with a negotiated sum from both bridge companies.

In the summer on 1848 work began on the construction of the Wheeling suspension bridge.

Point of Honor is the mansion where Elriva Daniel and Charles Ellet, Jr. were married in Lynchburg, Va., in the fall of 1837. She was the daughter of a wealthy judge and known for her beauty, graciousness and sterling character. Her devoted husband called her "Ellie," and when they were apart, they corresponded almost daily. The 1806 home is listed on the National Register of Historic Places and is open to the public. LYNCHBURG MUSEUM SYSTEMS

Details of French suspension bridges by Joseph Vicat, engineer of the Corps des Ponts et Chaussées. Cables, deck stiffening truss and towers are illustrated. The wire cables in the garland system are brought over the towers and anchored on the back to prevent deflection waves from passing between spans. ANNALES DES PONTS ET CHAUSSÉES, PARIS 1831

Fribourg Bridge, completed in 1834 and spanning 870 feet, was the longest bridge in Europe throughout the 19th century. Although located in Switzerland, it was built by the French engineer Joseph Chaley. SMITHSONIAN INSTITUTION

The cables of the Moussac Bridge, 1841, display the French garland system of hanging the vertical suspenders from separate main cables. The ligatures (cable wrappings) so common in French suspension bridges are clearly seen. They are an alternative to continuous wrapping with small diameter wire as featured on the Wheeling Suspension Bridge. KEMP COLLECTION

The Mallemort Bridge over the Durance River, circa 1846, incorporates many of the engineering details espoused by Joseph Vicat. KEMP COLLECTION VIA SMITHSONIAN INSTITUTION

Ellet's use of the French garland system is evident in this photograph of the Fairmount Bridge. SMITHSONIAN INSTITUTION

Ellet's first triumph as a suspension bridge builder was the Fairmount Bridge over the Schuylkill River at Philadelphia, 1842. The photo shows the well-known Fairmount Waterworks in the background as well as a horse-drawn streetcar and wagon on the bridge. FREE LIBRARY OF PHILADELPHIA

Building the National Road

*T*he National Road provided a strong incentive to bridge the Ohio River at Wheeling, and thus extend the road into Ohio and beyond. The Virginia Board of Public Works, established in 1816, oversaw the development of turnpike roads in the commonwealth as well as canals and other public works. In the same year, the legislatures of Virginia and Ohio incorporated a company to build a bridge over the Ohio, and went as far as appointing commissioners to carry out the act. By 1818, the National Road had reached Wheeling and was finally completed in 1821.

By 1808, the United States government was interested in internal improvements. One of their projects was to construct a great national road to connect the eastern seaboard with the vast resources of the western region. This first federal interstate would strike the Ohio River at Wheeling. In 1816, as construction was inching its way to Wheeling, eleven men obtained a charter from Virginia to build a toll bridge across the Ohio River at Wheeling. The name of the company was the Wheeling and Belmont Bridge Company, because a bridge across the Ohio River would connect Wheeling with Belmont County, Ohio. MUSUEMS OF OGLEBAY INSTITUTE

The "paved" road was completed to Wheeling in 1818. The man with his dog is facing Wheeling from the Ohio side of the river. Wheeling Island is to his left, and in back of the town is the National Road. MUSEUMS OF OGLEBAY INSTITUTE

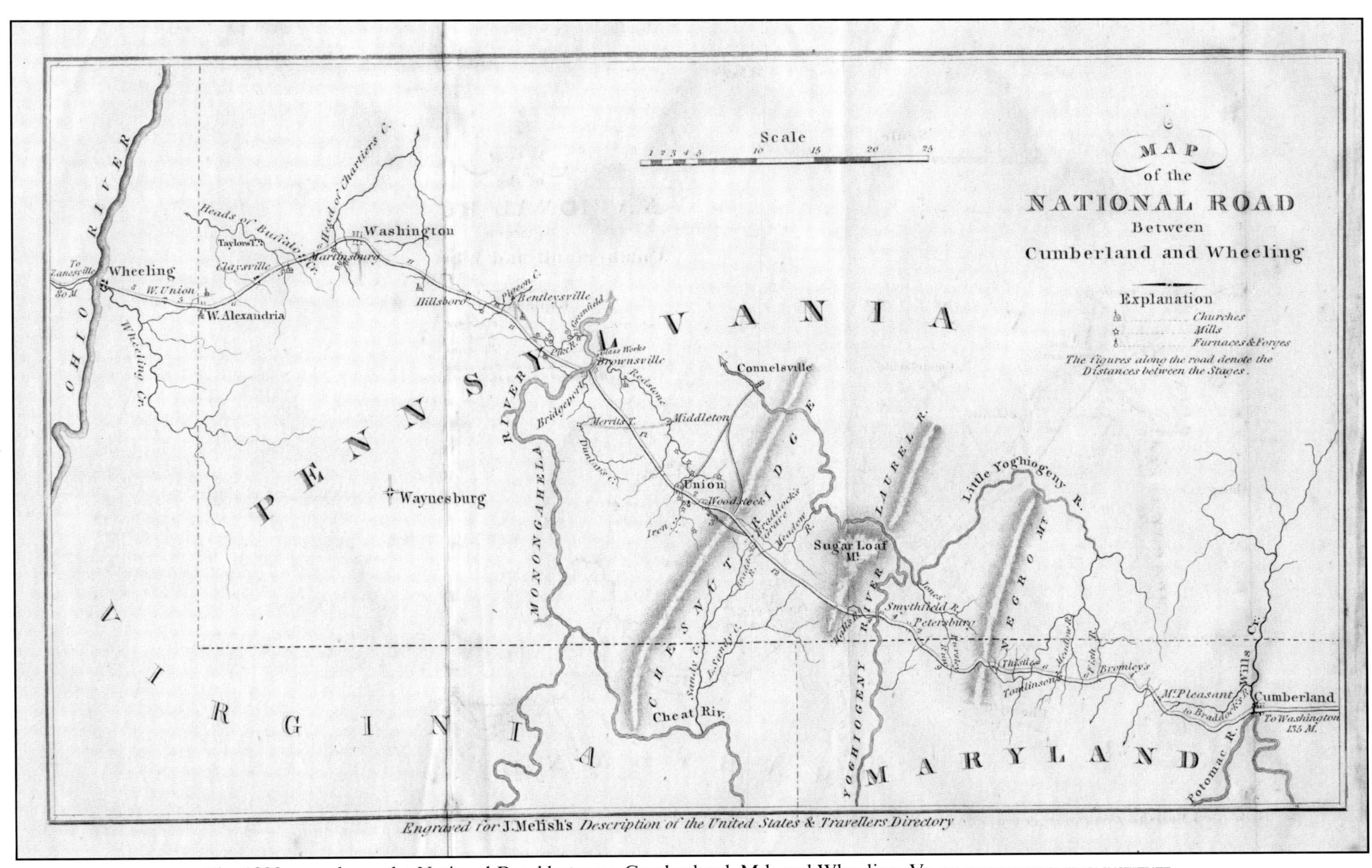

The 1822 map shows the National Road between Cumberland, Md. and Wheeling, Va. MUSEUMS OF OGLEBAY INSTITUTE

There were major Wheeling accomplishments in 1836. The Virginia General Assembly granted Wheeling a city charter, and the bridge company finally organized as specified in the 1816 charter. They were unsuccessful in persuading the federal government to provide bridge construction funds, so they raised the money locally in one month. They then proceeded to build the handsome covered bridge that same year across the back channel from the Island to the Ohio shore. MUSEUMS OF OGLEBAY INSTITUTE

The federal contract to build bridges in Virginia for the National Road was awarded to Moses Shepherd. The work did not include a bridge across the Virginia-owned Ohio River. It is believed that his wife, Lydia, was responsible for her husband building two unnecessary bridges across the creek in order that the road would pass right by their mansion. The circa 1817 bridge shown still carries heavy traffic near the 1798 home. The second bridge for Lydia was an interesting "S" bridge which has been replaced. Traffic through Elm Grove still crosses the creek two times with the extra bridges. MUSEUMS OF OGLEBAY INSTITUTE

This original National Road bridge with superb stone work is located near Blaine, Ohio. The bridge is no longer in use and is "endangered." MUSEUMS OF OGLEBAY INSTITUTE

This photograph of an "S" bridge in Ohio clearly illustrates the unusual configuration which looks like a letter S. By crossing a stream perpendicular to the channel, the masons avoided complex skew masonry arches which were probably beyond the skills of local contractors. MUSEUMS OF OGLEBAY INSTITUTE

The Wheeling Suspension Bridge

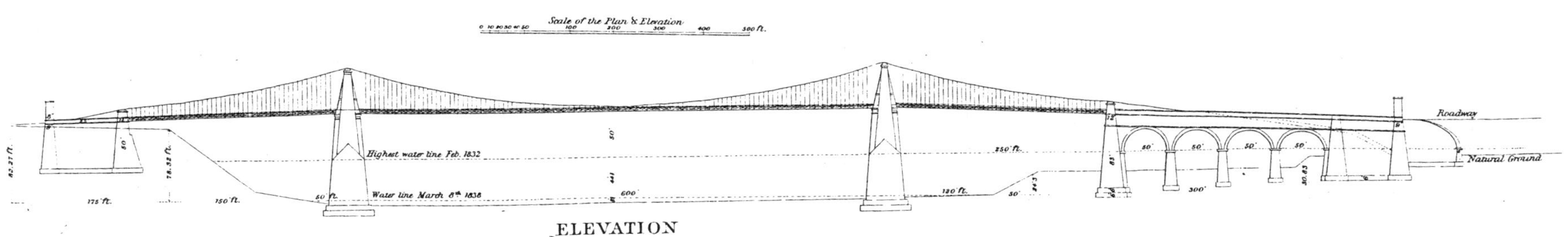

The Virginia Assembly reorganized the Wheeling and Belmont Bridge Company in 1836 through an act passed in March of that year, which was affirmed by the legislature in Ohio in 1837. One of the first activities was to construct a picturesque covered bridge composed of two carriageways and two sidewalks across the back channel of the Ohio River. During its lifetime, this bridge carried horse-drawn streetcars, as well as other traffic, before it was replaced by a handsome iron truss bridge fabricated by the Wrought Iron Bridge Company and erected in 1893, built on piers and abutments of the original covered bridge.

In 1847, the Wheeling and Belmont Bridge Company was reorganized with an authorization to issue $200,000 in stock, which was sufficient money to erect a wire suspension bridge. The authorization stated the bridge should not obstruct steamboat navigation on the river or it would be treated as a public nuisance and abated accordingly. However, the act did not specify the clearance required over the water to avoid a public nuisance, and this led to a long and acrimonious debate with the city of Pittsburgh.

When completed, the Wheeling suspension bridge with a main span of 1,010 feet was the world's longest clear-span bridge. It held this position of leadership until the completion of the Brooklyn Bridge in 1883. The Wheeling Bridge represents the beginning of America's ascendancy in long-span suspension bridges, which was to last more than a century before other nations completed long-span bridges to rival their American counterparts. The Wheeling Bridge was under construction from 1847 until the autumn of 1849. The principal features are massive stone towers to support the cables. As originally designed, Ellet used six separate cables on each side of the bridge to support the timber deck with light handrails. The bridge is not symmetrical, rising over the main channel, which runs along the Wheeling bank of the river, and descending nearly 40 feet to the island side.

All of the materials for the bridge were produced in Wheeling. The stonework was quarried downstream and brought to the site by barge. The timber was locally produced, as was the natural cement used in the mortar for the towers and wingwalls. The cement was also used to imbed the eye bars in the anchorages on both the island and Wheeling town sides. Although the stonework and timber deck were within the

Lts. Dutton and Sanders produced a design for a suspension bridge at Wheeling as required by the House of Representatives' order of 23 February 1838. This design, based upon Telford's Menai Bridge, was to be a monumental structure in the British tradition.

25TH CONGRESS, 2ND SESSION, REPORT 993 H OF R, 1838

ELEVATION

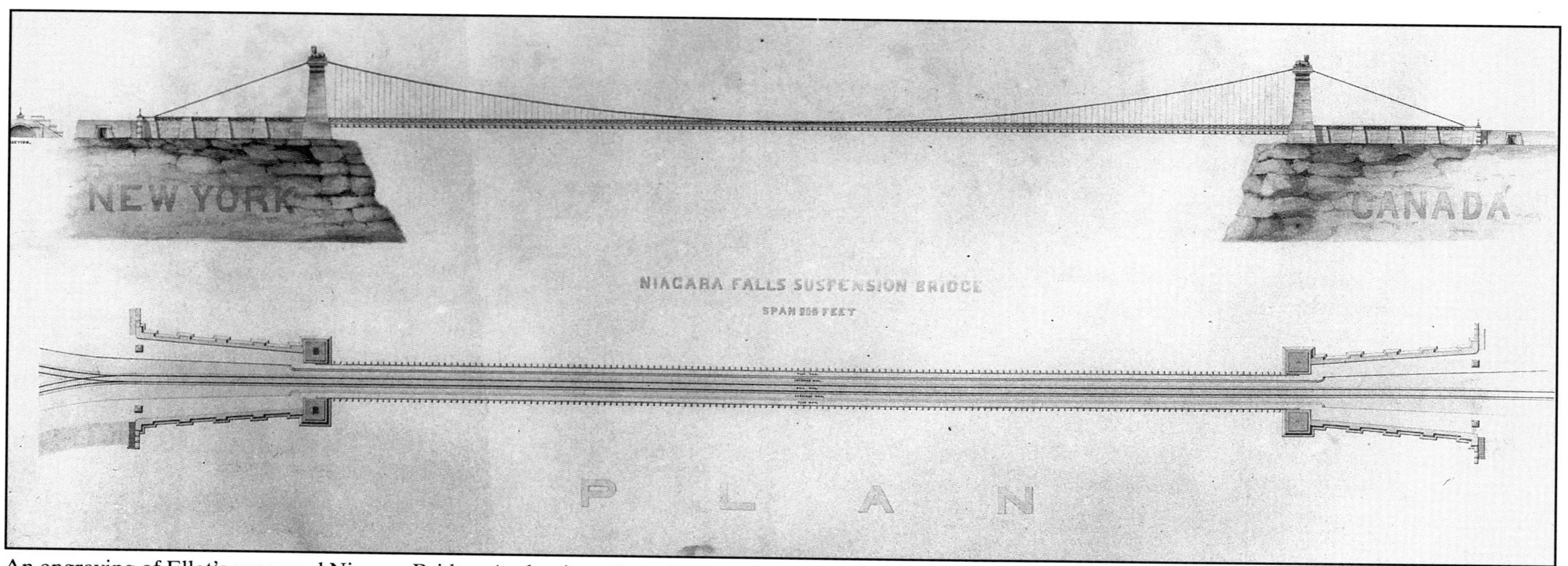

An engraving of Ellet's proposed Niagara Bridge. At the time, Egyptian Revival architecture was in vogue for engineering structure, and Ellet used Egyptian-inspired columns supporting the cables and surmounted with pairs of spinxes on each tower. This elegant bridge was never built since Ellet was relieved of the responsibility after the construction of the temporary bridge. KEMP COLLECTION

realm of traditional building, the use of wire cables was a totally new technology in the Upper Ohio Valley. David Richards and Josiah Bodley and Company furnished the wire, which was drawn in Wheeling. Through the act of drawing the wire through small dies, the strength is more than doubled and, equally important, each inch of the wire is tested as it is pulled through the die. If there is a defect, the wire will break. When used in a suspension bridge the full potential of the wire in tension can be utilized since the cables are in tension from anchorage to tower across the main span and to the anchorage on the other side.

The cables were made up on the island side, and one by one lifted into place on the bridge. With the cables in place, vertical suspenders could be attached to the cables and provided the means of support for the floor beams, which in turn supported the deck. Thus the bridge could be quickly constructed with no formwork necessary to support its construction.

Although the bridge company had voted to have a clearance over the main channel of 90 feet, while construction was well under way, the issue of clearance reared its head again. Pittsburgh interests claimed that it restricted navigation since the stacks of the largest steamboats would not clear the bridge, despite the fact that hinged stacks were fairly commonplace in river transportation at the time. The real issue here was for ascendancy of industrial and transportation power between Pittsburgh and Wheeling. Until the completion of the Davis Lock and Dam in the 1880s there wasn't a Port of Pittsburgh, and navigation above Wheeling was problematical, at best, during the summer boating season. The last plank of the floor system was laid on October 20, 1849, and the bridge was officially opened on November 15th to great public acclaim. As completed, the Wheeling Bridge featured 12 main cables, six on each side, utilizing the French garland system. Each cable was composed of 550 parallel #10 wires. The clear span was 1,010 feet. The original deck was only 960 feet, since it stopped short of the Wheeling or eastern tower to permit Water Street to pass in front of the tower and form a T-junction with the deck. The total dead load was estimated at 593,400 pounds. The magnitude of the structure amazed all that saw it directly as well as those familiar with

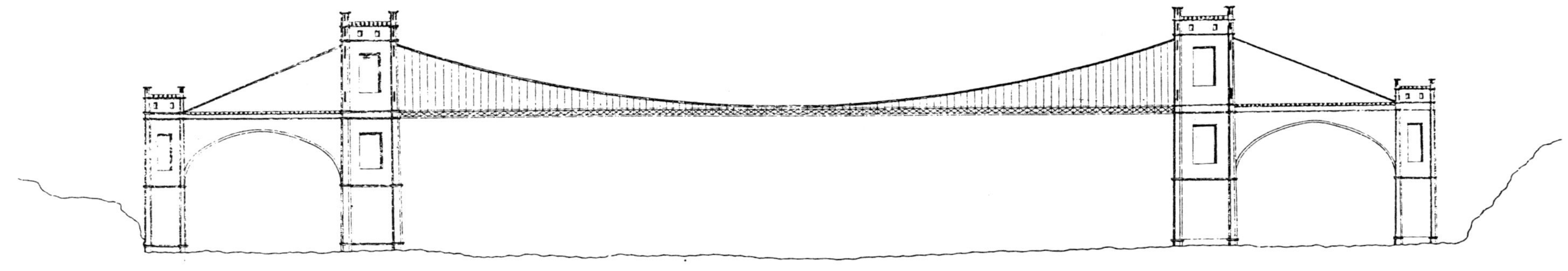

Above: Ellet's 1836 design for the Wheeling Suspension Bridge consisting of a 500-foot main span flanked by 100-foot arch side spans. At this time Ellet, writing from Lynchburg, Va., had not been to Wheeling. 25TH CONGRESS, 2ND SESSION, REPORT 993 H OF R, 1838

of...... *Josiah Fox* ...... Forty ......
first instalment of one dollar per share on...... Forty ......
al stock of the **Wheeling and Belmont Bridge Company**
l for by *A. C. Slutzker* adm ......

Wheeling, *12 April* 1847.
Wm. Paxton,
Thomas Sweeney,
Thomas Hughes,
Daniel Zane,
Wm. T. Selby,
James Baker,
Henry Moore,
Jas. E. Wharton,
D. C. List,

By *O. Morgan*
Treasurer.

Josiah Fox, the famous designer of the frigate, the U.S.S. *Constitution*, purchased stock shortly after the drawing at right appeared in the newspaper. The 1847 stock certificate includes a list of the "manager" of the bridge company. Thomas Sweeney was the president, who retained that position until 1875. Today he is better known for the huge glass "Sweeney Punch Bowl" made by Messrs. Sweeney & Co., located at the Oglebay Institute Mansion Glass Museum. E.C. JEPSON FAMILY

Below: The April 1, 1847, Wheeling newspaper included a sketch titled "The Bridge Over the Ohio" with a caption which read, "Above will be found a plain drawing of a Wire Suspension Bridge. We insert it that some, who have never seen one, may understand something of the nature of the structure." OHIO COUNTY PUBLIC LIBRARY

one or more of the many engravings produced in leading magazines and journals. It was touted as one of the great wonders of the modern age. It was, after all, the largest clear-span bridge in the world. The bridge was also an immediate financial success through its tolls, which enabled the bridge company to maintain the bridge in excellent condition for nearly 100 years.

Little significance happened between the completion of the bridge and 1854. This suddenly changed on May 17th when a devastating tornado moved up the Ohio Valley and destroyed the deck of the bridge.

This was not the first nor the last suspension bridge damaged by wind induced aerodynamic instability. Just as spectacular was the total destruction of the Tacoma Narrows Bridge in late 1940; in fact, the description of the collapse of the Wheeling suspension bridge could easily be applied to the Tacoma failure. The earliest recorded collapse of a modern suspension bridge was Samuel Brown's Brighton chain brige in 1833, which destroyed the superstructure. It was rebuilt, and destroyed a second time in 1836. During the same period severe damage was sustained by Telford's famous Menai Straits Bridge shortly after its completion in 1826, later in 1836, and again in 1839. Marching troops also caused several spectacular failures; the Broughton Bridge near Manchester, England in 1831 and a French suspension bridge at Angers. The French failure was ironic since marching troops were being used to load test the bridge when it failed. This disrupted and demoralized the French, which ended in an edict stopping further construction of such bridges in the country for the rest

Ellet's design of the Wheeling Suspension Bridge at the time of construction in 1847. No final drawings are known to exist. ELLET, *REPORT OF THE WHEELING AND BELMOUNT SUSPENSION BRIDGE,* PHILADELPHIA, 1847

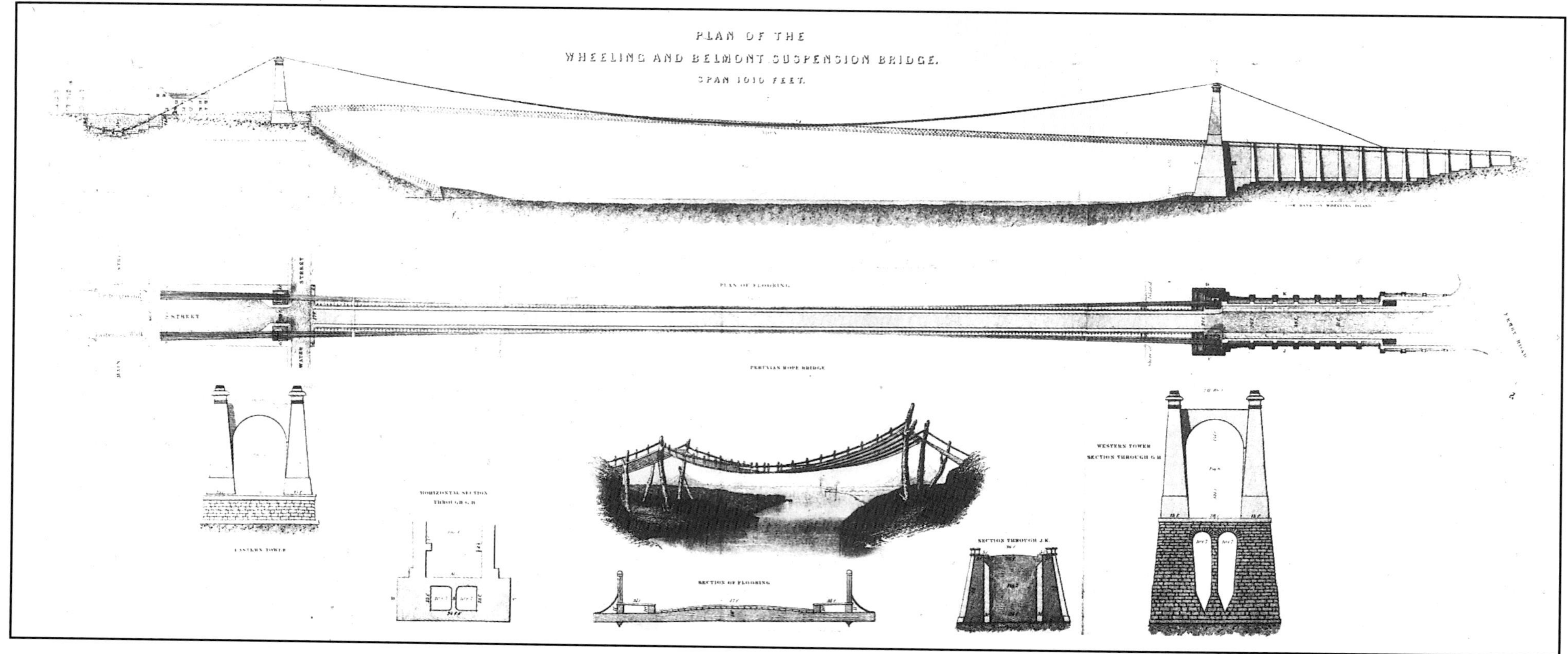

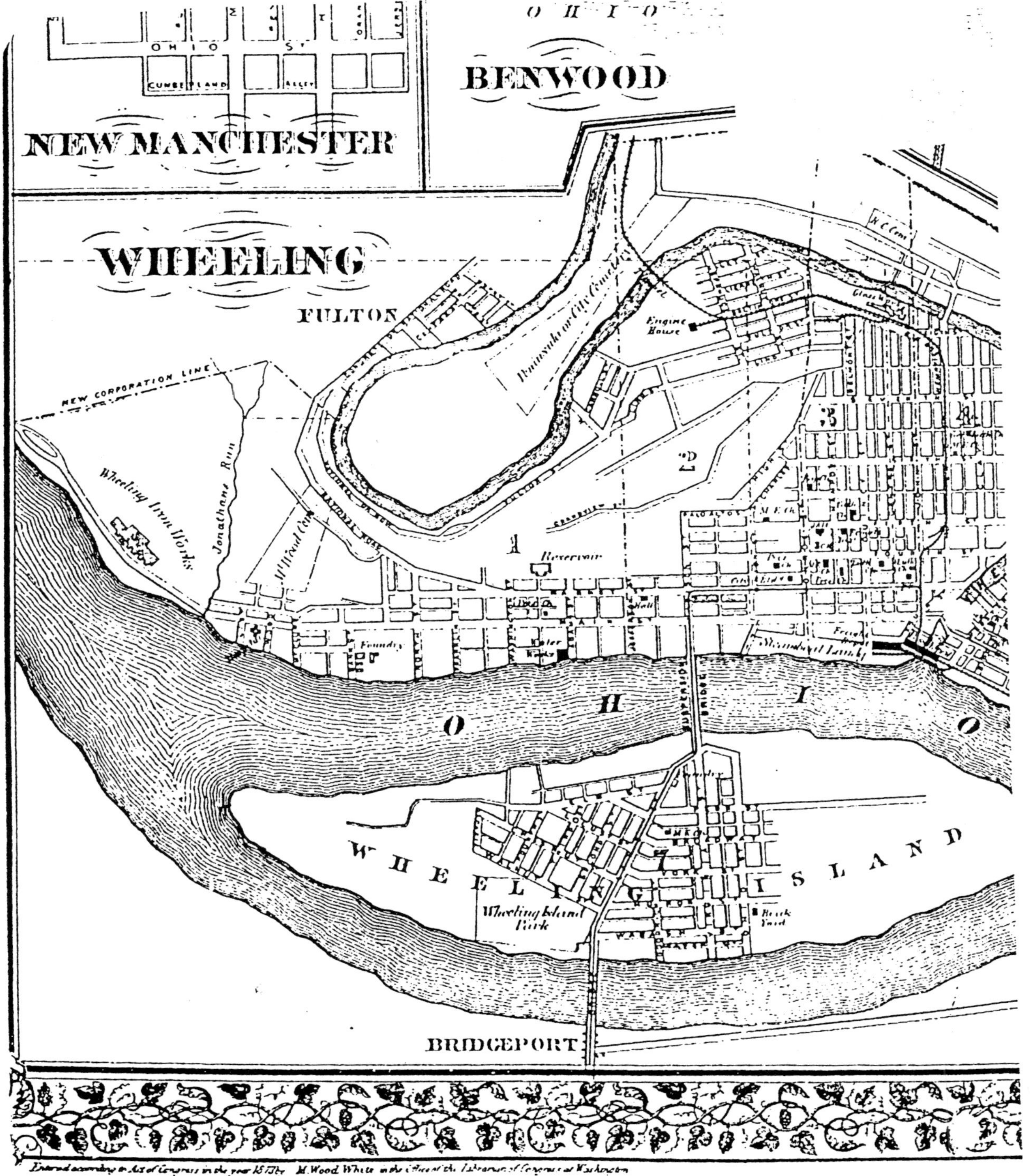

A map showing the location of the suspension and back channel covered bridges as well as the National Road and the center of Wheeling.

WHITE'S NEW COUNTY...ATLAS, GRAFTON, W.VA., 1873

of the 19th century.

There is a persistent legend that John Roebling came to the rescue, like a knight on a white horse, to rebuild the Wheeling Bridge. It was, however, Ellet who was called back to Wheeling to undertake the reconstruction of the bridge with his assistant, William K. McComas. In August, the Board of Directors commended McComas for a job well done in rebuilding the suspension bridge on a one-lane basis. He was later engaged to overhaul the bridge, which was completed in 1860. It was then that the cables were grouped into two pairs, one on each side of the bridge. Josiah Bodley supplied the new wire, while Thomas Sweeney and Company, well-known for their cast iron work in Wheeling, provided the castings necessary for the new bridge. Sweeney also served on the Board of Directors and as President of the bridge company.

Even before the collapse of the bridge, Ellet had discussed the possibility of putting streetcar traffic on the bridge. Shortly after the failure, Ellet indicated that it was an opportunity to place track on the deck to provide public transportation to Wheeling Island and to Ohio. It wasn't until 1867 that the Citizen Railway Company was authorized to lay tracks on the bridge for horse-drawn streetcars.

John A. Roebling, Sons & Company were the leading wire rope manufacturers in the country, utilizing the technique of wire spinning, which was patented by John A. Roebling, based on an earlier precedent by

Joseph Vicat in France. The bridge superintendent, Joseph Lawson, met with Washington Roebling and engaged him to prepare a report on the Wheeling bridge. It was this involvement of a Roebling that gave credence to the legend that John A. Roebling rebuilt the bridge. It should be noted that Washington Roebling did the work after the death of his father, and there is no evidence that he actually came to Wheeling. The design essentially Roeblingized the bridge with the diagonal cable stays that are such a prominent feature of the bridge, and the renewal of the stiffening truss of the McComas reconstruction. The rehabilitation work according to the Roebling design was completed in 1872. Although not a prominent part of the structure, Joseph Lawson was authorized to install ten gas lamps on the bridge as a matter of public safety. The only other illuminations previously were the 1,000 lights representing each foot of the bridge featured in the opening celebrations in 1849. In 1898, these gas lamps were replaced with ten arc and six incandescent lamps, greatly improving the illumination, while at the same time reducing cost to the bridge company.

Wilhelm Hildenbrand was one of the assistants to Washington Roebling on the construction of the Brooklyn Bridge. He was engaged shortly after that bridge was completed to produce a design for installing a new stiffening truss and moving the main cables laterally to accommodate a wider deck. It should be noted that Hildenbrand was the designer of the steel bridge, which was completed just downstream from the suspension bridge, and the back channel truss bridge, which replaced the original covered bridge. He was also responsible for the design of a large steel truss system on the Roebling Cincinnati Bridge in order to accommodate rail traffic. In a move to ensure the safety and durability of the bridge, the bridge company involved Professor Thomas Lawson of Rensselaer Polytechnic Institute in Troy, New York, to undertake regular inspections of the bridge beginning in 1913. By the end of the 1920s, the bridge company was interested in overhauling the bridge and strengthening it to a design by Professor Lawson, which would have greatly altered its appearance with very large steel trusses. The Great Depression hit and revenues from tolls dropped precipitously so that the company was in no position to undertake this major work. It was at this time that a number of interested parties contacted the bridge company about buying its

An engraving showing the original Ellet garland system with six wire cables on each side of the deck. ELI BOWEN, *RAMBLES IN THE PATH OF THE STEAM HORSE*, PHILADELPHIA, 1855

entire property, which included the suspension bridge, the connecting road, and the back channel bridge.

Citizens up and down the Ohio River remember the devastating St. Patrick's Day flood in 1936. Virtually all of Wheeling Island was under water, and the bridge was closed for two days, but the damage sustained to the bridge was minimal, involving mainly the storm cables that were put in earlier and descended from the deck to anchorages on the river banks. The repairs were in the neighborhood of $2,000. Another modification occurred in 1937, when sheet piling was driven in front of the island tower to prevent scour at high water stages of the river. It was not clear why this work was undertaken, but it may well have resulted from the laying of a sewer line in front of the tower on the island bank of the river. At the same time, the State Road Commission of West Virginia was pressing its case a second time for acquiring the bridge. After more than a decade of various suitors wishing to consummate a purchase of the bridge, it was sold to the city of

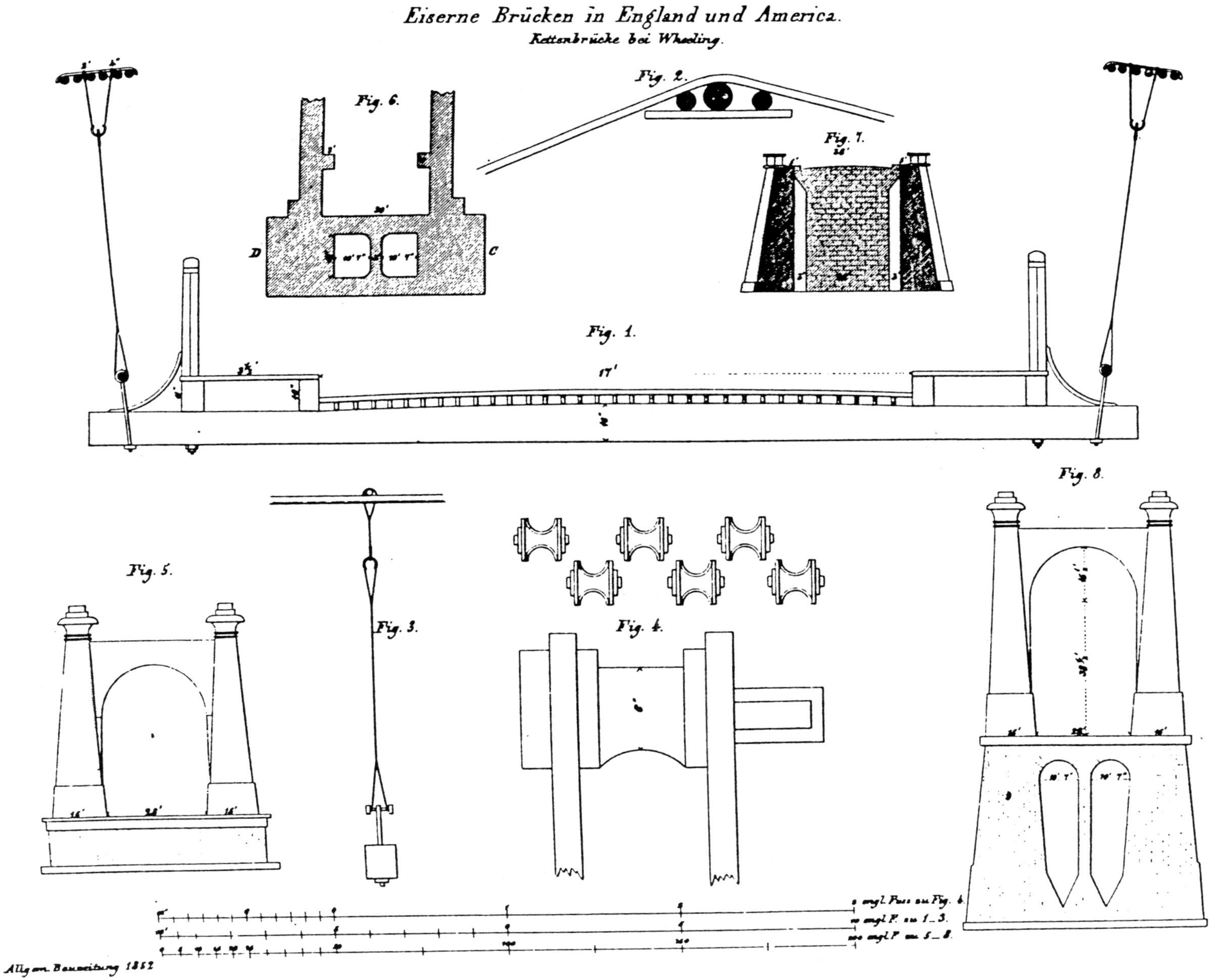

Details of the Wheeling Bridge. WIEN, FORSTER'S *ALGEMEINE BARZEITUNG* 1852, PLATE 485

From the beginning, Pittsburgh had objected to a bridge being built across the Ohio River at Wheeling. During the summer of 1849, the Pittsburgh newspaper articles became vehement. There was even an unsuccessful attempt to have a mob formed to blow up the bridge. Shortly thereafter, there was an appalling surprise. On July 28th, the Bridge Company was notified of a pending suit. Nine days later they learned that United States Supreme Court Justice Robert C. Grier was to hear an application for an injunction from the Commonwealth of Pennsylvania against the bridge company. It was claimed the bridge would be an obstruction to navigation on the Ohio River. It was evident Pennsylvania was representing the interests of Pittsburgh. After hearing arguments from representatives of both Pennsylvania and Wheeling, Justice Greer reached a decision on August 30. The case would be brought before the entire United States Supreme Court. At the time the bridge construction was proceeding rapidly, and by October 4th the last main cable was swung across the river, and each cable was being wrapped with wire as a protection from the atmosphere. Seven days later the flooring was almost ready to be crossed by heavy vehicles. On Saturday, October 20, 1849, the newspaper announced that "the structure may be used to a limited extent, after to-day." People were most curious about this bridge which was so high in the air. They arrived early the following morning "thronging both sides of the river." At ten o'clock a 30-star American flag was hung on the city tower, and the Ohio state flag was flying on the western tower. Engineer Charles Ellet, Jr. and I. Dickinson, Superintendent of the stone and iron work, drove onto the bridge in a one-horse carriage as the last deck timber was swung into place and covered with planks. The carriage continued to the western shore as the roar of cannon announced their safe arrival. There was a "long, triumphant shout" from the "thousands of delighted spectators. In the afternoon, the vehicle of the world famous circus performer, General Tom Thumb, was driven across the bridge. The carriage and two Shetland ponies were given to him by Queen Victoria of England. Tom Thumb, the diminutive eleven year old who never grew to be more than thirty-six inches tall, finally arrived in Wheeling a few days later for his performances. He said he was very sorry to have missed the occasion of the first crossing.

Wheeling for $2.15 million in 1941. Thanks to the careful maintenance of Lawson and his successors, the bridge was found to be in excellent condition. The *Wheeling Intelligencer*, two years later, noted that West Virginia Governor Neely was present in Wheeling to accept from the City of Wheeling the suspension bridge, the back channel bridge, the steel bridge and the Etnaville Bridge as part of the state highway system.

The bridge remained in an essentially unaltered appearance for more than a century until 1956 when a new steel deck system was installed consisting of steel floor beams and an open steel deck. It compromised the historic appearance of the bridge, but from an engineering point of view it was a decided advantage, as this deck would spill the wind through the deck in a high-wind situation and greatly improve its aerodynamic behavior. This deck system has been retained through the renovation work completed in 1983 and the major overhaul of the bridge in 1999.

In 1969, the American Society of Civil Engineers conferred landmark status on the bridge, followed in 1975 by the National Parks Service giving the bridge national significance. Despite these designations, the *Wheeling News Register* reported in June 1979 that replacing the bridge was an option being considered. But it was later stated that based upon its historic significance and the generally sound condition of the bridge, it was decided to repair rather than replace this internationally significant structure. It was during the 1983 repair that the consultants, Howard, Needles, Tammen and Bergendoff specified the use of Neoprene cable wrapping as an economy measure and touted its durability. It was their opinion that, because the cables were closely spaced, it would be impossible to re-wrap them in the time-honored wire wrap system. It is interesting to note that at the same time the Roebling Aqueduct at Lackawaxen, completed in 1847 with pairs of cables, was successfully re-wrapped. A leading consultant noted that in his opinion, wire-wrapped cables are the Cadillac job. The wire keeps the strands tightly compacted to prevent intrusion of water; gives with the movement of the cables and provides maximum protection against damage to the main wires. Equally important, its appearance is far superior to any other system yet devised.

Because of sag, the Neoprene wrap was not durable and permitted the ingress of water with notable corrosion of the wires, thus

MONROE HOUSE.

W. BARRETT, Proprietor.

EAST END OF THE WIRE SUSSENSION BRIDGE,

Wheeling, Virginia.

Despite the legal difficulties, a grand celebration was planned for the formal opening of the longest bridge in the world. Wheeling City Council provided $100 for the occasion, and a committee was formed of members of City Council and the Bridge Company. They set the celebration date to be November 15, 1849, and mailed invitations. The memorable event began early that day with the roar of cannons. Soon there were people everywhere. The bridge was lined with vehicles, and a band from Zanesville supplied the music. All day there were crowds on the bridge until 3:00 when the ladies, who were attired in their best finery, and their escorts promenaded over the bridge to the Island and back. At six o'clock, 1,000 lamps were hung on the wires, and they were lighted almost simultaneously. "They presented an elegant and graceful curve of fire, high above the river, that was never excelled in beauty." It reminded people of the remark Senator Henry Clay had said while looking at the bridge and thinking of the impending court case. He remarked, *Take that down? You might as well try to take down the rainbow.*" There were the usual speeches which took place on a rostrum in front of the Monroe House on Main Street. This was followed by a "supper" at the Monroe House for which tickets had been sold. There were all sorts of food served such as oysters, chickens, ducks, "sallads" and more. "Sparking sherry and lively champagne" were served, while there were speeches, letters read from dignitaries and many, many toasts. The President of the Baltimore and Ohio Railroad Company reminded guests that the arrangements were being made to place the road under contract to the eastern terminus of Wheeling. The once in a lifetime celebration ended late in the evening. WEST VIRGINIA COLLECTION, WEST VIRGINIA UNIVERSITY

neccessitating the 1999 restoration of the bridge. This included the repair of the stiffening truss, selected suspenders, and wire rope stays, and a complete inspection and re-wrapping of the main cables. The 1983 project was completed for $2.5 million, whereas the present project to a design by A.G. Lichtenstein and Associates and undertaken by American Bridge Company, is in excess of $7 million.

The bridge, as we see it following the 1999 renovation, has all the essence of the original Ellet design, but it has been modified over the years, particularly with regard to the deck system and the installation of the stay cables. Thus, we view the bridge as a continuum from 1849 to its 150th anniversary in 1999. The bridge is one of the world's outstanding historic bridges and one of the greatest of the antebellum American civil engineering structures.

One of the earliest photographs of the Wheeling Suspension Bridge during the flood of 1852. The garland cables are clearly visible although hidden in a shadow. The original anchorage housing can be seen in the foreground together with the tollhouse.

1850-1860: Court Case and the 1854 Wind Storm

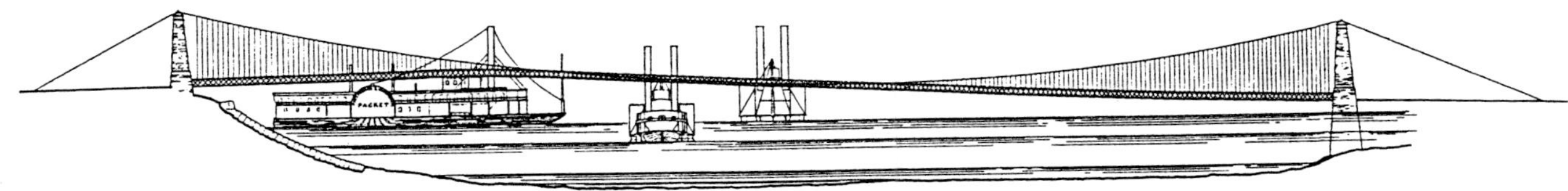

On November 10th, 1849, five days before the formal opening of the Suspension Bridge, the water level in the Ohio River rose to 20 feet at the site of the bridge. The steamboat, *Messenger*, shown in the illustration, was unable to pass under the bridge until Captain William Dillon cut down seven and a half feet from her chimneys. Captain Dillon was employed as the Suspension Bridge toll collector. The Pittsburgh newspaper accused the bridge company of ordering Dillon to cut the stacks, while the Wheeling newspaper assured its readers that the Captain of the *Messenger* had enlisted his help. The following day, when the water level was even higher, the *Hibernia #2*, on the way to Pittsburgh, was unable to pass under the bridge. The captain of the *Hibernia #2* moved his passengers and cargo to other conveyances and then waited for the water level to recede to continue the journey upstream. Other steamboats which were able to lower their chimneys had no problem passing under the bridge. Pennsylvania filed another supplemental bill stating the obstruction by the bridge to the *Messenger* and the *Hibernia #2* caused "loss and injury" to the owners, the trade of Pittsburgh and the commerce of Pennsylvania. COLLECTION OF THE PUBLIC LIBRARY OF CINCINNATI AND HAMILTON COUNTY

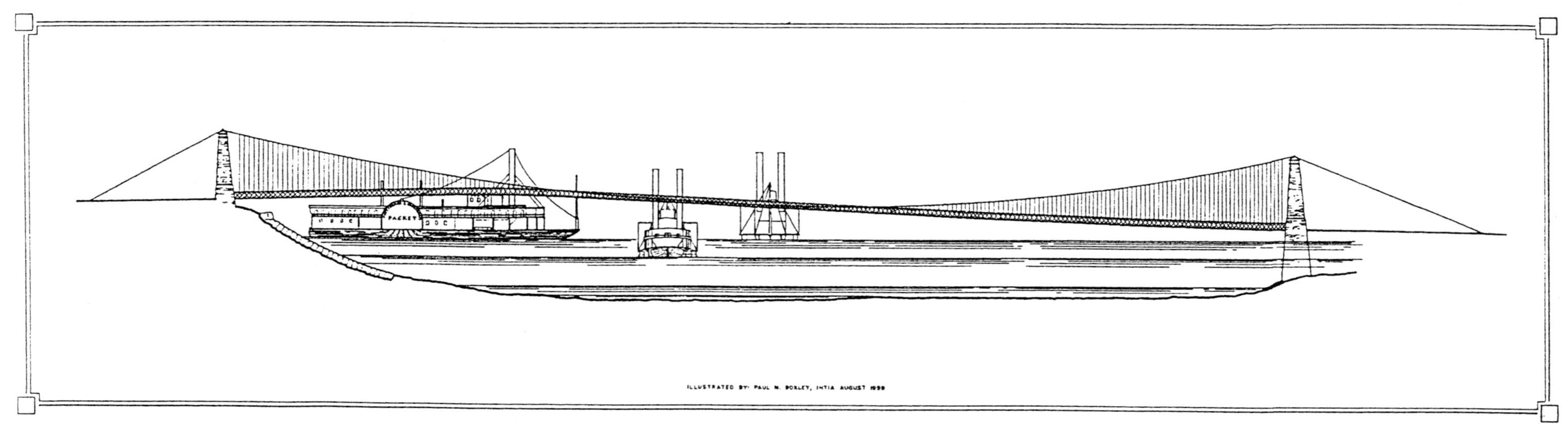

In May of 1850, the case was presented to the U.S. Supreme Court. The court referred the case to Reuben Walworth to take further "proofs" to determine if the bridge was an obstruction to navigation on the Ohio River. A year later, on May 13, 1851, Commissioner Walworth presented his 770 page report. It stated the bridge is an obstruction to boats with tall chimneys. The court found in favor of Pittsburgh. This drawing was one of the exhibits. It shows that steamboats which were unable to lower their chimneys could not clear the deck during periods of high water. RE-DRAWN BY PAUL BOXLEY OF THE INSTITUTE FOR THE HISTORY OF TECHNOLOGY AND INDUSTRIAL ARCHAEOLOGY FROM THE ORIGINAL AT THE NATIONAL RECORD CENTER

There was an advertisement in the 1851 City Directory for the Eagle Wire Mills. It included a drawing of the suspension bridge, and the ad stated that they make wire for bridges. That statement leads to the assumption that they made the cable wires for the bridge in Wheeling. This is incorrect as "the wire for this stupendous and beautiful structure" was made by Joshua Bodley and David Richards. MUSEUMS OF OGLEBAY INSTITUTE

This 1851 *Directory of the City of Wheeling & Ohio County* has an engraving of "the world-wide celebrity," the Wheeling Suspension Bridge. There was also a long description of the legal case and these bridge statistics: *The western tower column is 60 feet high while the tower on the Island is 69-3/4 feet; the length of the structure is 1,010 feet with 12 iron cables, 10 large and 2 small which support the flooring; the larger cables each contain 140 strands of number 10 wire while the smaller cables have 50 strands; the cables are anchored by a succession of links which look like a huge chain; the cables on the Wheeling side are anchored in massive masonry walls under Main Street while those on the Island are anchored in the wing walls which extend from the abutment on the Island; the flooring is attached to the cables by "wire stays" 3/4 of an inch in diameter, varying in length as they approach and recede from the towers; the highest elevation of the flooring is over the channel of the river, 212 feet from the Wheeling shore and then the flooring descends to the Island and the bridge would hold an army of 4,000 men.* The toll rates in 1849 were for both bridges to Bridgeport, Ohio, or to the Island and back: *Foot passengers—5 cents; a man and horse—10 cents; two horse carriage or wagon—20 cents; 6 horse wagon—75 cents; each horse or mule—8 cents; and hogs and sheep were 2 cents each. There were also tickets available such as "foot passenger ticket for a family per year" at a cost of $10.00.* MUSEUMS OF OGLEBAY INSTITUTE

1852 was a year to remember in the annals of Wheeling. On February 6th, the high court delivered its opinion. The bridge was an obstruction to steamboat traffic and had to be removed or elevated to be 111 feet above low water. There were further options to consider such as having a "draw" in either the Suspension Bridge or the Back Channel covered bridge. On April 30, there was a devastating 38-foot flood. It inundated Wheeling Island and large portions of the city. The daguerroeotype clearly identifies the eastern tower of the bridge with the original twelve cables, anchorage housing and the tollhouse. In the lower right hand corner is a stage stop with vehicles for National Road travelers. MUSEUMS OF OGLEBAY INSTITUTE

WHEELING WAS BEGINNING TO RECOVER FROM THE TRAGIC APRIL 1852 FLOOD WHEN DEVASTATING NEWS ARRIVED IN WHEELING. THE SUPREME COURT ANNOUNCED THEIR FINAL DECREE IN LATE MAY. THE WHEELING SUSPENSION BRIDGE HAD TO BE ELEVATED TO BE 111 FEET ABOVE LOW WATER FOR A LEVEL SPAN OF 300 FEET OVER THE CHANNEL OR REMOVED BY FEBRUARY 1, 1853. TO SAVE "WHEELING'S RAINBOW" IT WOULD HAVE TO BE HEIGHTENED 21 FEET IN LESS THAN NINE MONTHS.

The Bridge Company's next move was to have the bridge "legalized." They appealed to the Committee on Post Offices and Post Roads in the House of Representatives. The proposal was to have the Suspension Bridge and the Back Channel Bridge designated as post and military roads. A bill was drafted, but it was lost in the rush at the end of the session, so a rider was attached to the post office appropriation for the bridge. It passed both houses of Congress during the last three days of the session and was signed by the President. It was approved on August 31, 1852. The August legislation also stated all vessels were required to regulate the use of the vessels and of their "chimnies" in order that they did not interfere with the bridge. The steamboat *Dick Fulton* illustrates how the stacks were lowered. The long and costly legal battle was over and it seemed the bridge was safe at last. The year ended well for the city on Christmas Eve when the Baltimore & Ohio Railroad tracks were joined at Rosbey's Rock in Marshall County, Virginia. The momentous event marked the beginning of national rail transportation from Baltimore to the Ohio River. In January 1853, the first train arrived in Wheeling.

On the evening of May 17, 1854, a newspaper writer was on the Wheeling Suspension Bridge when it began to sway violently. He hurriedly left the bridge and shortly thereafter witnessed the collapse of the structure. The next day his detailed account of the "violent storm" which blew down the bridge was published in the newspaper. "At last there seemed to be a determined twist along the entire span, about one half of the flooring being nearly reversed, and down went the immense structure from its dizzy height to the stream below, with an appalling crash and roar." This 1854 newspaper article matches the Tacoma Narrows Bridge failure which occurred in the State of Washington on November 7, 1940. The description of the death throes of the Wheeling Suspension Bridge is consistent with these two Tacoma Bridge photographs. Engineers learned about the aerodynamic behavior of suspension bridges from the Tacoma failure. SMITHSONIAN INSTITUTION

The Wheeling-built steamboat, the *T. Swann,* was named for the president of the famous Baltimore and Ohio Railroad Company. It was one of the seven steamboats which carried passengers and cargo to and from the Wheeling B&O Railroad depot in Wheeling. *RAMBLES IN THE PATH OF THE STEAM HORSE,* 365

Below: The watercolorist Lefevre J. Cranstone depicts the bridge shortly after the temporary repairs were completed. MUSEUMS OF OGLEBAY INSTITUTE

The Board notified Charles Ellet, Jr. of the bridge failure, and he immediately came to Wheeling. Judge Grier signed another injunction against Ellet and the bridge company. It was to halt the rebuilding of the bridge, unless it was heightened to 111 feet, as required in the 1852 Supreme Court decree. Engineer Ellet voiced his concern that the high court might not agree that Congress had the right to regulate commerce among the States. In spite of his reservations, using salvaged cables, he rebuilt it to the original height in just forty days. It was only 10 feet wide, with the "carriage way" six feet and two "footways" of two feet each. There was constant communication between watchmen on each end to prevent collisions. It is interesting that many of the 1849 cable wires which were salvaged from the bottom of the Ohio River are still in use today.
MUSEUMS OF OGLEBAY INSTITUTE

Once again there was a United States Supreme Court case. The defense stated the Congressional Act had legalized the bridge at a given height, therefore, the 1852 decree could no longer be enforced. The Court agreed and dissolved the injunction in 1856. That decision was the end of the lengthy and complicated legal case. There were almost one hundred witnesses. The separation of power among the three branches of the United States government, as well as states rights, were difficult issues during the case. The bridge company had won "the monstrous attempt of a portion of the people of Pennsylvania to destroy one of the grandest works of the age." The entire bridge litigation took place in the United States Capitol. This circa 1846 photograph shows the Capitol before the 1851 construction began for the two wings to be used by the Senate and the House of Representatives. U.S. SENATE HISTORICAL OFFICE

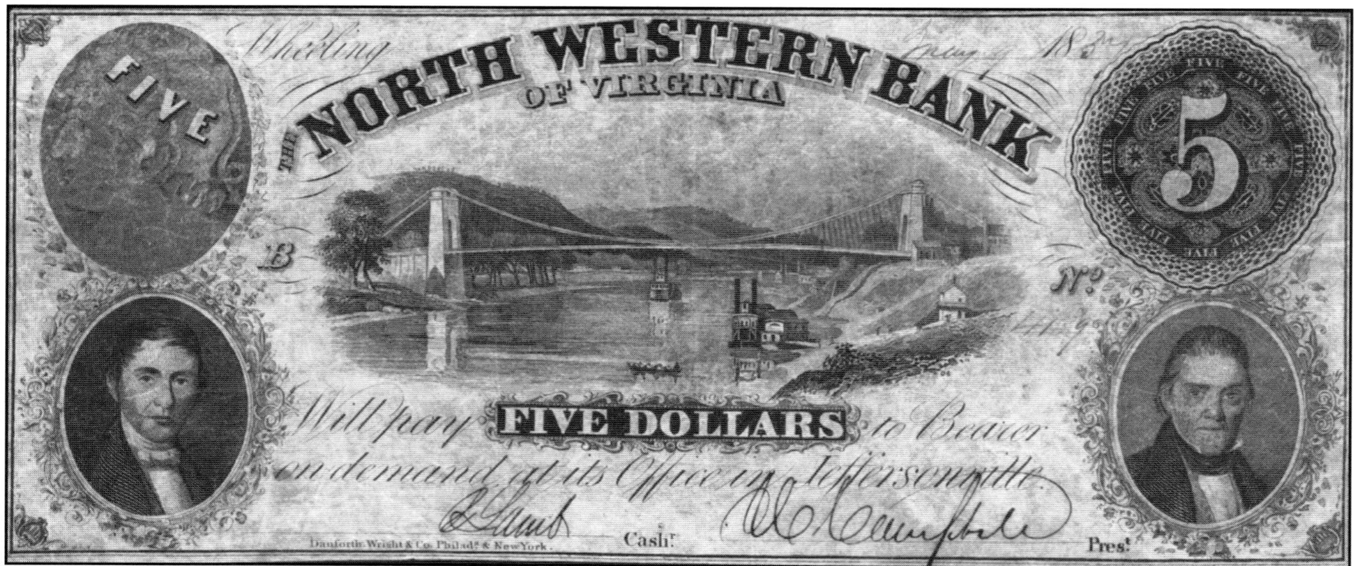

The costs incurred as a result of the bridge legal case caused severe money difficulties for the Bridge Company. On July 19, 1859, the managers agreed to suspend the dividend and to borrow $20,000 from the North Western Bank to rebuild the bridge. The lineal descendent of this bank, chartered in 1817, is Bank One. R. MICHAEL BAKER

At the time this engraving was made, circa 1856, Wheeling was the crossroads of America. The Ohio River provided transportation for people and goods traveling north and south or east and west. The city was a United States Port of Delivery. The Baltimore and Ohio Railroad provided fast transport from Baltimore to Wheeling, and the National Road, with the bridges over the Ohio, had reached Vandalia, Illinois. OHIO COUNTY PUBLIC LIBRARY

Two female slaves belonging to John Hunter escaped into Ohio by crossing the bridge on the evening of November 25, 1850. The officer who chased them was unable to catch them and returned empty-handed. They were the first slaves to flee from Wheeling after the passage of the fugitive slave law. MUSEUM OF OGLEBAY INSTITUTE

The *New York Times* on June 2, 1860 published the "View of Wheeling." The description of the industrial city stated the sky was not canopied with smoke as it was in Pittsburgh, but there was enough shiny soot to darken brick buildings, damage furniture and blacken carpets. By that date, the bridge construction was well under way with many of the "Emerald Island" men working on the cables at a "giddy altitude." A newspaper article stated that the cables were to be painted white, the footwalk banisters red and the suspenders blue. MUSEUM OF OGLEBAY INSTITUTE

On August 1, 1860, the bridge rebuilding
was complete, and people no longer had
to use the ferry, wade or straddle a log to
reach the opposite side of the river. The
cables were combined to two on each side
and inclined to the vertical to increase the
lateral stiffness of the bridge. In the
vertical dimension the heavy stiffening
trusses indicate the lesson learned from
the 1854 storm. However, although it was
again full width, the foot-paths were on
the outside of the cables. This arrange-
ment was a "little inconvenient" where the
foot-paths were partially obstructed by the
cables. Obviously all the people in the
photograph decided it was easier to walk
on the roadbed. The scuppers, openings
under the walkways, were used to dispose
of animal waste and water. This is the only
information known on the McComas/Ellet
rebuilding of the bridge. CREATIVE IMPRES-
SIONS STUDIO

Civil War and West Virginia Statehood

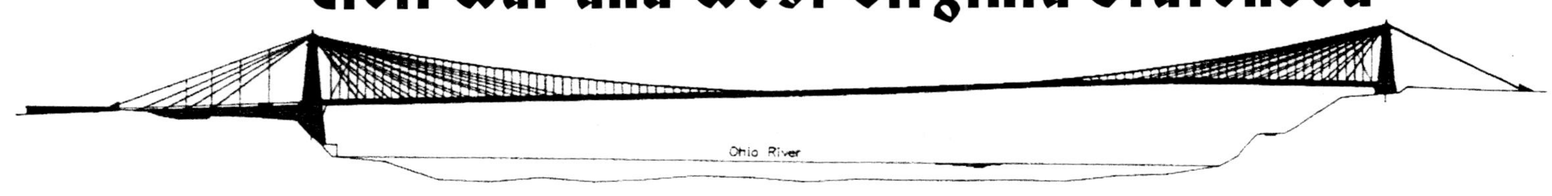

When a "crack" company of immigrant German riflemen embarked for active duty in steamboats in 1861, Wheeling was already involved in the Civil War. The majority of people were for the Union, while those who favored the South were in a difficult situation. Many of them went south, and some men joined the Confederate army. Everyone had to take an oath of loyalty to the government of Virginia which had cast its fortune with the Union. If they refused, they might be sent to the Atheneum prison which was nicknamed "Lincoln's Bastille." It was located across the street from the Custom House. The federal court room in the Custom House was used for treason trials for those people who were suspected of being traitors to the United States. The streets were filled with soldiers who were often unruly. A permit was required to leave the city. Business flourished with the manufacture of all kinds of war materials from ambulances to ammunition, but it was not a pleasant time to live in the city. MUSUEMS OF OGLEBAY INSTITUTE

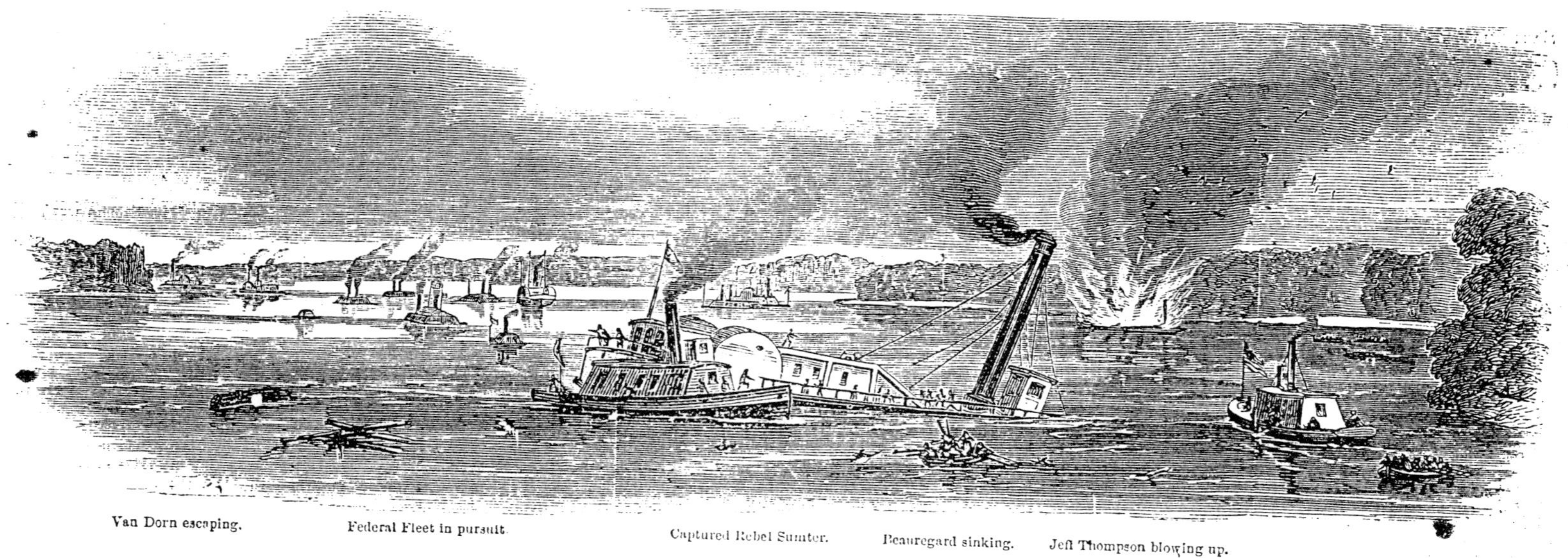

Charles Ellet, Jr. was called to meet with Edward Stanton, the Secretary of War, following the successful attack of the *Virginia* at Hampton Roads. The Confederate ram ship had sunk two ships and killed over two hundred Union men on March 8, 1862. Stanton commissioned Ellet, Colonel of Engineers. He was authorized to build and man a fleet of steam rams and execute a counter attack. Colonel Ellet outfitted the best vessels he could find and selected his brother, Alfred, to be second in command. On June 6, 1862, his ram fleet encountered the enemy on the Mississippi River, near Memphis, Tennessee. They sank four ships and scored a great victory. His son, Charles Rivers Ellet, who was with him on the flagship, *Queen of the West*, accepted the surrender of Memphis. Colonel Ellet was the only Union casualty. He was "slightly" wounded while viewing the action. KEMP AND FLUTY COLLECTIONS

Two weeks after the surrender of Memphis, Colonel Ellet died. His body
was taken to Independence Hall in Philadelphia where it lay in state under
the Liberty Bell. Male relatives and friends were invited to attend the
funeral. Carriages were available to take them to the cemetery. When it was
time for the funeral procession, the coffin was removed from Independence
Hall and placed in a "splendid plumed hearse." The long line of carriages
to Laurel Hill Cemetery was accompanied by the slow movement of martial
music. In attendance were members of the family, members of councils,
naval and military officers and a military escort. The famous engineer has
not been forgotten in Wheeling. On June 26, 1999, the 150th year of the
bridge, members of the Wheeling Area Historical Society placed a bronze
plaque on Ellet's monument which reads:

Charles Ellet, Jr.

Jan. 1, 1810 — June 21, 1862

Eminent Civil Engineer,

Canal, Railroad,

Suspension Bridge Builder—

Wheeling, 1849

INDEPENDENCE NATIONAL HISTORICAL PARK

During the Civil War there was an encampment of Union soldiers, such as
John Prager from Wheeling, on Wheeling Island at Camp Carlile. At times
there were as many as 10,000 men stationed there. The soldiers often
caused problems after consuming too many alcoholic beverages. A city
ordinance helped to correct this annoyance by forbidding merchants to
serve such beverages to them. In addition, guards were stationed on the
bridge to enforce the regulation that no soldiers were allowed to leave the
Island after six o'clock in the evening. Tolls for the soldiers use of the bridge
were to be paid by the federal government. The Bridge Company had a
difficult time collecting these funds. After several trips to Washington and
continuing entreaties, they finally received $3,885.70 in January of 1871.

MUSEUMS OF OGLEBAY INSTITUTE

The people of the western counties of Virginia had been generally antagonistic to the rest of the state, and there was even talk of separation. When Virginia joined the Confederacy in 1861, the people in this region voted to remain loyal to the United States government. They organized the Restored Government of Virginia and fought on the side of the Union. The United States Constitution has a provision to create a new state from an existing one. The Restored Government of Virginia used this measure to create a new state from Virginia. They followed the legal steps required, but the final decision rested with the United States government. The West Virginia statehood bill was approved by the House of Representatives and the Senate. It was signed by President Abraham Lincoln only a few hours before it would have failed, on December 31, 1862. West Virginia officially became the 35th State in the Union on June 20, 1863. The Wheeling Custom House, a federal building, served as the Union Capitol of Virginia from 1861 until West Virginia statehood. It is considered to be the birthplace of the State. This National Historic Landmark, now named West Virginia Independence Hall, is a state museum. FLUTY COLLECTION

THE CUSTOM HOUSE AT WHEELING, VA., NOW THE SEAT OF THE NEW GOVERNMENT OF VIRGINIA.—FROM A SKETCH BY OUR SPECIAL ARTIST.

Wheeling: 1870-1900

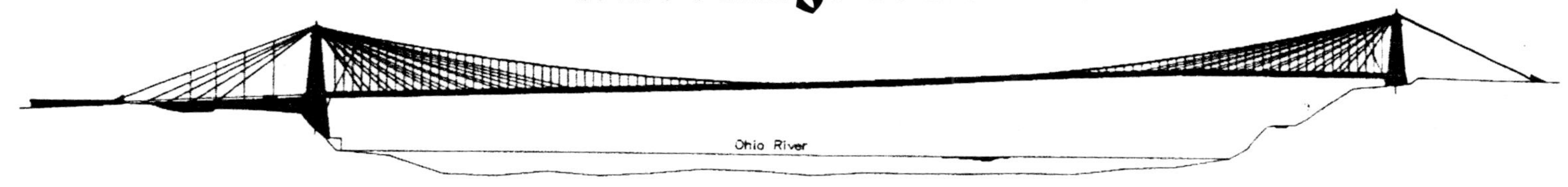

The 1853 Baltimore and Ohio Railroad Freight Shed and the 1873 Depot were located at the junction of Wheeling Creek and the Ohio River. After unsuccessful efforts to save the historically important structures, the Wheeling Civic Center was built in that location in 1977. A Fink truss from the Freight Shed was removed and sent to the Smithsonian Institution. LIBRARY OF CONGRESS

Railroad tracks passed by the entrance to the Back Channel Bridge at Bridgeport, Ohio, just as they do today. The photograph is not identified as to what caused the people to look so intently at the river.

This enlarged portion of a circa 1885 photograph reveals the splendid buildings on Main Street. The tracks on the bridge were used for the "street railway" which was horse-drawn. On the Island, the photograph clearly shows an area used for boat launching. OHIO COUNTY PUBLIC LIBRARY, BROWN COLLECTION

Thomas Edison demonstrated the electric lamp in 1879, and three years later Wheeling became the fifth city in the nation to have electric lights. In 1885, two lamps were installed on the bridge, obviously making travel easier and safer. Electric and telephone lines were strung across the bridge providing service to the Island.

An 1889 photograph of the waterfront shows crowds standing on the Suspension Bridge and many people on the waterfront. A yacht had capsized, and three people drowned. OHIO COUNTY PUBLIC LIBRARY, BROWN COLLECTION

The *Lizzie Townsend* is shown icebound at the Wheeling Wharf about 1896. This happened each year when the Ohio River was frozen solid. Wheeling was often the "head of navigation" in the winter with the ice and during the summer when the river was low. The local newspaper provided the depth of the river daily during the summer, named the steamboats arriving and departing from the wharf, and often listed the cargoes they transported. MUSEUMS OF OGLEBAY INSTITUTE

Wheeling was booming when this photograph was taken around 1895. Public buildings, homes, factories, warehouses and railroads are all shown. The bridges and buildings at the mouth of Wheeling Creek were being used by the B&O railroad. The large cylindrical structure in the lower right hand corner is the "gasometer." Next to the Suspension Bridge is the 1891 steel bridge. MUSEUMS OF OGLEBAY INSTITUTE

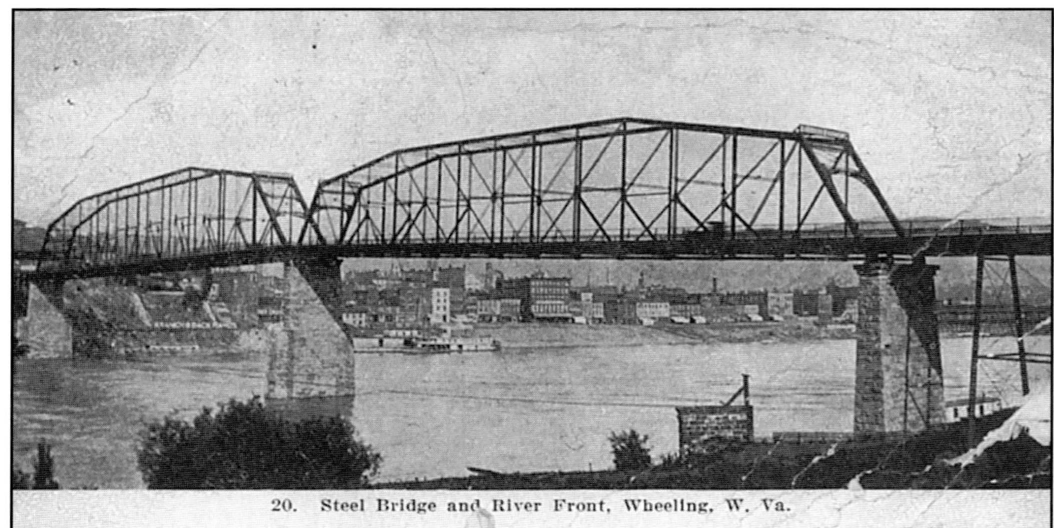

20. Steel Bridge and River Front, Wheeling, W. Va.

The steel bridge was erected just downstream from the suspension bridge and served both road and rail traffic to the island. It was designed by Wilhelm Hildenbrand, who was a leading engineer on the Brooklyn Bridge in 1883, as well as adding steel stiffening trusses to strengthen the Cincinnati Suspension completed by John Roebling in 1867. ELLEN DUNABLE, WHEELING

Flying fragments of fire from three burning buildings in Bridgeport, Ohio, landed on the wooden roof of the Back Channel Bridge and set it on fire. A telephone call was made to the Wheeling Fire Department, who quickly responded. A nozzleman laid a hose on the roof of the bridge but was forced to leave his position because of the "fierce and blinding smoke." Another fireman took his place and kept working while the toll keeper poured water over him, or he "literally would have been roasted." He was badly burned, but his heroics and the work of others on May 10, 1883, saved the bridge. OHIO COUNTY PUBLIC LIBRARY, BROWN COLLECTION

Wheeling and Its Bridge During the 20ᵗʰ Century

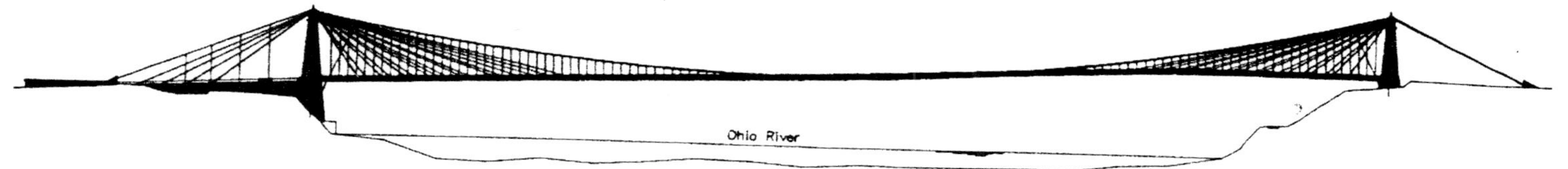

Unfortunately it is not known if Joseph Lawson is included in the photograph titled "Lawson's Crew." He was the superintendent of the Wheeling and Belmont Bridge Company. Mr. Lawson's total commitment to excellence kept the Suspension and Back Channel bridges in superb condition. MUSEUMS OF OGLEBAY INSTITUTE

At right: The school house floating near the
Suspension Bridge started its March journey
during the flood of 1907 in Warenton, Ohio,
and ended its over 100 mile trip at Sistersville,
West Virginia. FLUTY COLLECTION

Above: The details of the Suspension Bridge on
this May 1913 photograph show well, along with
the marine railway, the Mull Center then called
the Hawley Building, the Capitol Music Hall and
a train with many cars on Water Street. OHIO
COUNTY PUBLIC LIBRARY

At right: The policeman standing on the brick
pavement at the intersection of Main and 10th
streets is a reminder of West Virginia prohibition.
The state was officially "dry" at midnight on June
30, 1914. The Wheeling and Belmont bridges
provided easy access for illegal refreshments to be
brought into the city from Ohio. Five years later
there was national prohibition. MUSEUMS OF OGLEBAY
INSTITUTE

Photo by Prof. Thomas Lawson,
circa 1920s, showing the timber
deck, main cables, suspenders
and stay cables.

Wheeling has a long history of floods, but the worst on record was the flood in March of 1936. The river reached a height of 55.5 feet. It covered the Island and almost all of the downtown reaching to 16th Street and Chapline. Approximately 20,000 people were forced to evacuate their homes. The Island was isolated. The Suspension Bridge, the Steel Bridge and the Back Channel Bridge were closed. Tragically, lives were lost, but there was little damage to the Suspension Bridge. The storm cables had been purposely cut to avoid strain from the force of the rushing water. ZEE PHOTO, WHEELING

The Managers of the Bridge Company authorized $100 for this "ad" to be placed in the 1936 Wheeling City Centennial Celebration. Long before Wheeling was a city, the area in front of the Suspension Bridge was the site of the first market house. The small frame building with three or four stalls was a favorite site for the cows and hogs to meet at night after roaming through the town all day. Their grunts and bawling greatly disturbed the residents who were trying to sleep. MUSEUMS OF OGLEBAY INSTITUTE

Typical pedestrian ticket, used until the bridge was sold in 1941. KEMP COLLECTION

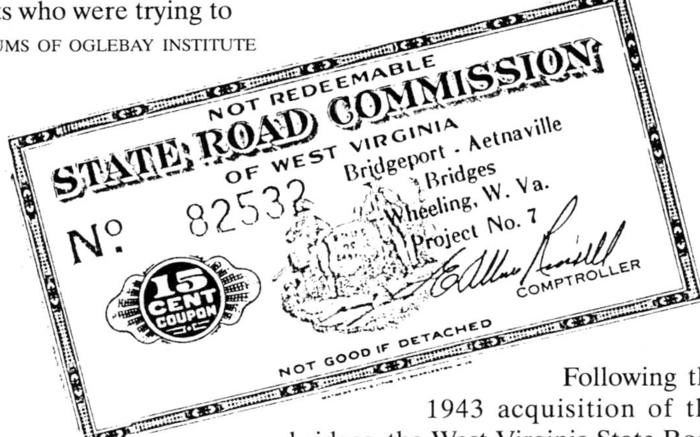

Following the 1943 acquisition of the bridges, the West Virginia State Road Commission issued books of toll tickets. They were used for the Bridgeport (Back Channel) bridge or the Aetnaville bridge. The books containing ten tickets was sold for $1.00, which saved the customer fifty cents. Those two bridges are now closed. A new bridge carries local traffic from the Island to Bridgeport, Ohio. ELIZABETH STEINBICKER FARIS

The arrival of the circus was a big event in Wheeling. Many people fondly remember the elephants walking across the Suspension Bridge. Some Wheeling residents say that the elephants also swam across the river. Perhaps they did, as the National Geographic Society says elephants do swim, but then again, perhaps they just walked across the river when the water was shallow. MRS. HUGH ERWIN

The 1955 Fort Henry Bridge is a tied arch steel bridge which serves Interstate I-70. The original Fort Henry was a federal fort that protected Wheeling during the Revolutionary War. Ironically, the approaches to the Fort Henry Bridge fit the 1851 City Directory description of one of the suggested federal plans to heighten the Suspension Bridge. It stated there would be "piers in the streets" and "streets running under the floor!!" WEST VIRGINIA DIVISION OF HIGHWAYS, DISTRICT 7

A major change in the appearance of the bridge occurred in 1956 when the wood deck and floor beams were replaced by an open grid steel deck. This modified deck provided greatly increased resistance to wind loads, being stronger and able to spill wind through the grid if there is any tendency for uplift on the deck. DEAN R.P. DAVIS, 1957

A dramatic view of the recently installed open grid steel deck. DEAN R.P. DAVIS, 1957

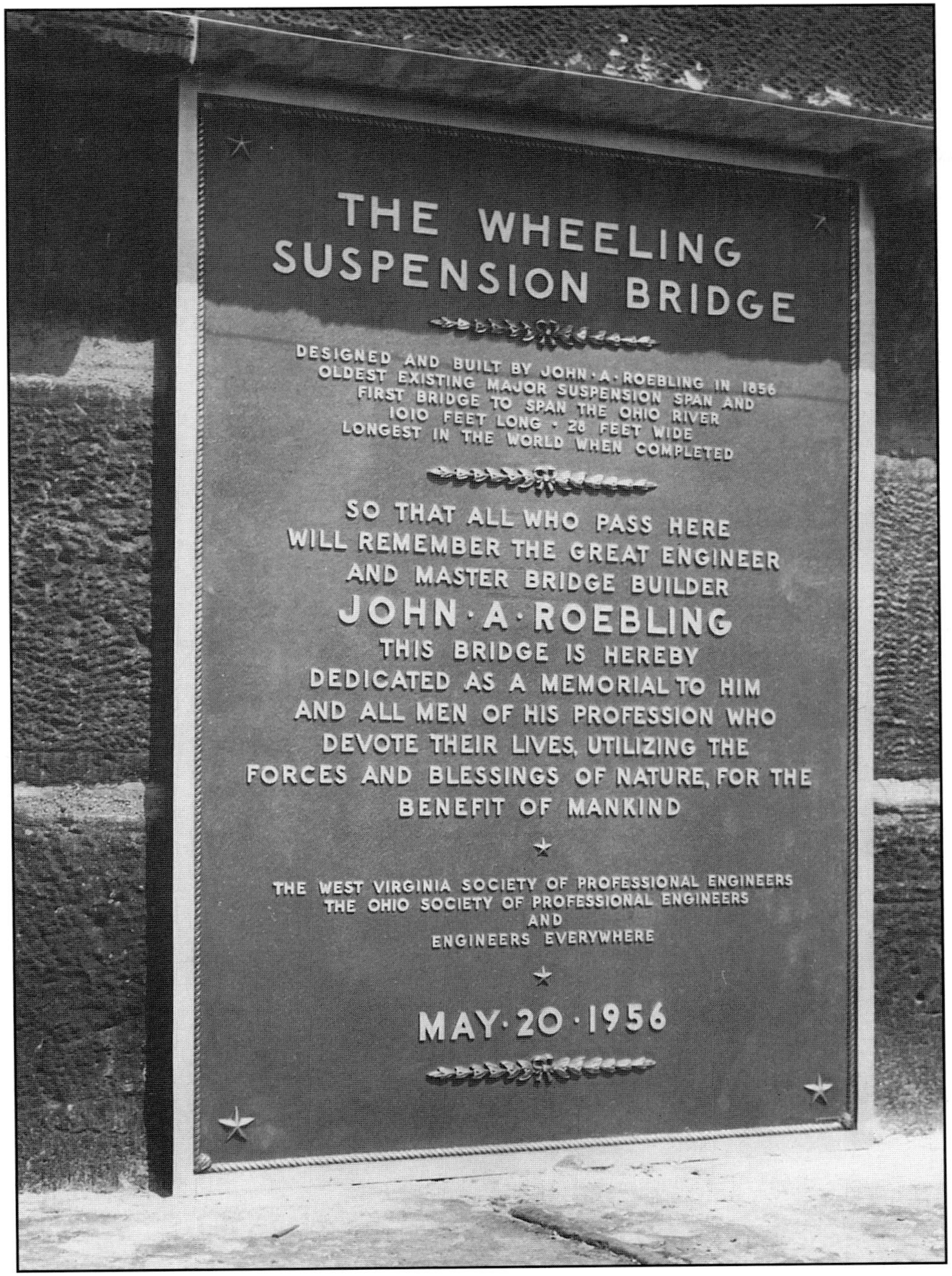

At left: On May 20, 1956, a Suspension Bridge plaque was placed on the west end of the bridge in the south wing wall. The speaker for the occasion presented the correct bridge history, however, the marker contained inaccurate information. It stated that the bridge was built by John A. Roebling and completed in 1856. The plaque was removed years later when the error was discovered. DEAN R.P. DAVIS

Below: On June 21, 1969, the American Society of Civil Engineers designated the Wheeling Suspension Bridge a National Historic Civil Engineering Landmark. This plaque is correct as it named Charles Ellet, Jr., as the engineer. In 1969, an inspection stated the bridge was safe but deteriorating. By 1976 when the bridge was designated a National Historic Landmark by the United States Department of the Interior, the deterioration was noticeable to even a casual observer. Friends of Wheeling, a local historic preservation organization, led the campaign to have the bridge repaired. FLUTY COLLECTION

AMERICAN SOCIETY
of
CIVIL ENGINEERS

designates

THE WHEELING SUSPENSION BRIDGE

as a

NATIONAL
HISTORIC CIVIL ENGINEERING LANDMARK

WHEELING, WEST VIRGINIA
JUNE 21, 1969

Although the individual cables, six on each side, were grouped into two pairs of cables, the original garland cables can be seen as they approach the anchorage. The later wire rope stay cables can also be seen in this same view as they approach the anchorage on the Wheeling side. WILLIAM E. BARRETT

The notches in the stringcourses on both towers are the result of the cables on the south side of the bridge being thrown on the arch during the violent storm of 1854 and are being retained as a reminder of this disastrous event. WILLIAM E. BARRETT

Details of the deck showing the timber Howe truss in 1976. WILLIAM E. BARRETT

This 1976 photo shows half the main span and the island tower. Inset shows the main and stay cables entering the Northeast anchorage. The massive anchorage vault is located under the street intersection. Buried in the vault are large iron "eye" bars, which extend beyond the face of the anchorage to receive the wire cables. WILLIAM E. BARRETT

An aerial view of the bridge in 1976 before the 1983 renovation. WILLIAM E. BARRETT

In 1979 there was a thorough inspection of the bridge. The report stated that the bridge could not continue to be used much longer unless there were major repairs. It also said that the cost to do the needed repairs could only be justified because of the historical significance of the structure. The bridge was designated an endangered National Historic Landmark by the U.S. Department of Interior and The National Trust for Historic Preservation. The State of West Virginia seriously considered replacing the suspension bridge with a new bridge. Local, state and national organizations joined forces to persuade officials to repair the bridge. Fortunately, they were successful, and a $2.4 million contract replaced the stiffening trusses, repaired the cables and anchorages, replaced the wire wrap on the main cables with neoprene and other work. On May 5, 1983, the official grand reopening was celebrated. One of the ten most significant bridges in the county had been saved. HOWARD, NEEDLES, TAMMEN AND BERGENDOFF

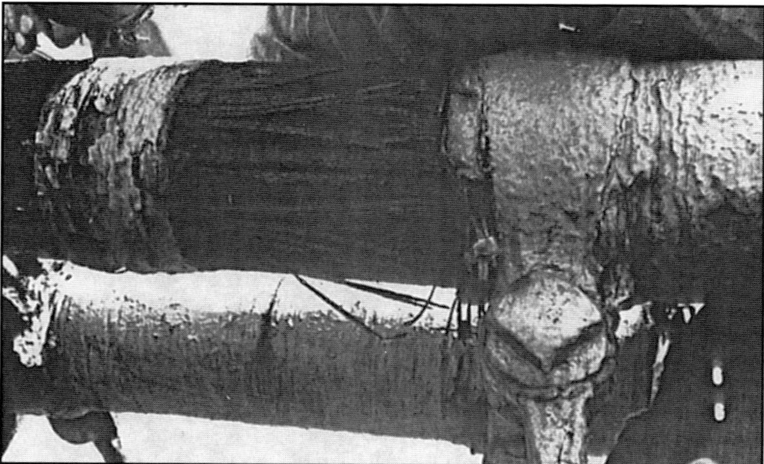

Repairing main cables and deck during the 1983 renovation. ZEE PHOTO, WHEELING

Close-up view of a main cable showing the parallel wires and the cable clamp that attaches the suspenders. The Neoprene wrap can be seen on the right. ZEE PHOTO, WHEELING

It was a special Thursday evening, on December 10, 1987, when a switch turned on the Suspension Bridge "necklace of lights." The Wheeling Convention and Visitors Bureau and Wheeling 2000 had raised the funds from the community. The lights were universally admired. Sadly, after several years, the lights were turned off due to a lack of money for maintenance and repairs. They were sorely missed, and a plan emerged. The Wheeling National Heritage Area Corporation (WNHAC), using federal funds, designed an elaborate lighting scheme which was installed in 1999. The future of the lighting is secure. Wheeling officials have pledged to pay for the electricity, and the State of West Virginia has committed to providing maintenance. DICK CRESS—ELLEN DUNABLE COLLECTION

The bridge repaired and reopened following the 1983 renovation.

A complete overhaul of the Wheeling Suspension Bridge was begun in 1998 and finished in time for the bridge's 150th anniversary in October 1999. American Bridge undertook the work to a design by A.G. Lichtenstein. Every effort was made to preserve the historic fabric while insuring the safety of the traveling public. This photograph illustrates the early stage of the bridge renovation with the installation of a safety net. ZEE PHOTO, WHEELING

Main cable clamp, dated 1860, with elastomeric wrap removed, showing parallel wires. ZEE PHOTO, WHEELING

Opening the main cables with wood wedges to inspect the wires. ZEE PHOTO, WHEELING

Cable wrapping machine designed by American Bridge for use on the Wheeling Suspension Bridge. ZEE PHOTO, WHEELING

Close-up of the wrapped cable with the 1860 cable clamps supporting the suspenders which, in turn, carry the deck. ZEE PHOTO, WHEELING

A view inside the anchorages showing the connection of the wire cable to "eye" bars protruding from the anchorages. Note the cast iron wedges behind the ladder used to tighten the cables. ZEE PHOTO, WHEELING

A typical stay cable fitting. ZEE PHOTO, WHEELING

Close-up of one of four finials mounted on top of the towers. These decorative elements serve as a means of protecting the saddles from the weather. ZEE PHOTO, WHEELING

Cleaning the timber truss prior to painting the wood Venetian red, an authentic mid-19th century color. A red (ferric oxide) paint was specified in 1860 when the bridge was rebuilt by W.K. McComas. The main and stay cables are white with Prussian blue suspenders to replicate white lead and Prussian blue of the original color scheme. ZEE PHOTO, WHEELING

Workmen removing old deck and installing a new self-rusting open grid deck. ZEE PHOTO, WHEELING

Wheeling Bridge Chronology

65-1706	As many as 24 iron chain link bridges erected in the Himalayas.
1535	First information on Inca bridges reached Europe.
1756	Birth of James Finley in Northern Ireland.
1801	First modern suspension bridge over Jacobs Creek in Western Pennsylvania by James Finley.
1806	Birth of John Augustus Roebling.
1808	Finley and Templeman suspension bridge patents.
1810	Birth of Charles Ellet, Jr. Finley's article on chain link suspension bridges appeared in *Port Folio*
1811	Samuel Brown proposed bars coupled together instead of link chains.
1816	Hazard and White erected a wire suspension bridge over the Schuylkill River, Philadelphia. Richard Lees completed a wire suspension bridge over Galawater, Galashiels, Scotland. Charter for the Wheeling and Belmont Bridge Co.
1818	National Road reached Wheeling.
1820	Samuel Brown's Union Bridge completed.
1823	Navier's report on suspension bridges. Marc Sequin's first bridge erected at Annonny, France.
1825	Tournon-Tain Bridge by Sequin over the Rhone.
1826	Telford's Menai Straits Bridge.
1827	Ellet's first job on a survey party on the North Branch of the Susquehanna River.
1829	Navier's ill fated Pont des Invalides collapsed.
1830	Ellet arrived in France.
1831	Following lectures at the Ecole des Ponts et Chaussées, Ellet surveyed suspension bridges in the Rhone Valley.
1832	Ellet submitted design for a suspension bridge across the Potomac.
1836	Ellet appointed Chief Engineer of the James River and Kanawha Canal Co. Ellet's first design for a suspension bridge at Wheeling.
1837	Ellet married Elvira Daniel.
1838	John A. Roebling wrote to Ellet about a position on the J.R.&K.C.
1839	Ellet published: *A Popular Notice of Wire Suspension Bridges.* Prepared detailed *Report on a Proposed Suspension Bridge Across the Mississippi.*
1840	Ellet promoted suspension bridges in the Ohio Valley and elsewhere.
1841	Ellet corresponded with John A. Roebling regarding wire cables.
1842	Fairmount Suspension Bridge completed.
1843	Ellet began contacts with Wheeling businessmen for a suspension bridge across the main channel of the Ohio River.
1845	Ellet began design for the Niagara Suspension Bridge.

1847   Ellet appointed Chief Engineer for the Wheeling and Belmont Bridge Company.
Appointed Chief Engineer for the Niagara Suspension Bridge.
Published: *Report on the Wheeling and Belmont Suspension Bridge.*

1848   Construction underway at both Niagara and Wheeling for suspension bridges.
At the end of the year Ellet resigned from the Niagara project after controversy with the two bridge companies.

1849   WHEELING SUSPENSION BRIDGE COMPLETED.
Lawsuits began by Pittsburgh interests to remove the bridge as an impediment to navigation.

1852   Ellet's second proposal for a bridge over the Potomac.

1854   Great storm destroyed deck and flings cables into the river. Bridge rebuilt by Ellet and Willam McComas.

1856   Roebling's double deck Niagara Suspension Bridge opened.

1860   Wheeling bridge rebuilt by McComas which significantly strengthened and altered the appearance of the bridge.

1862   Ellet commissioned Colonel in Command of ram fleet which he constructed.
Battle of Memphis—a Union victory but Ellet was wounded and died later, June 21.

1867   Citizen Railway Company was authorized to lay tracks on the bridge for horse-drawn street cars.

1872   Joseph Lawson, Superintendent for the Bridge Company completed the rehabilitation of the bridge according to a design by Washington Roebling. The design called for wire rope stay cables which "roeblingized" the appearance of the bridge.

1874   Gas lamps first installed on the bridge.

1883   Brooklyn Bridge completed.

1887   Wilhelm Hildenbrand, Assistant to Washington Roebling on Brooklyn Bridge, prepared a design for the bridge which was completed by Joseph Lawson and consisted of new timber stiffening trusses and moving the cables laterally to accommodate a wider deck.

1913   Professor Thomas Lawson appointed engineering consultant to the Bridge Company.

1923   Joseph Lawson died.

ca. 1929   Professor Lawson asked to design a large steel stiffening truss to increase both the strength and stiffness of the bridge. Because of the Great Depression the design was not implemented.

1936   Devastating St. Patrick's Day flood caused minimal damage to the bridge.

1938   Floor beams over railway line renewed.

1941   City of Wheeling purchased bridge for $2.15 million.

1943   Bridge transferred to state of West Virginia.

1956   Major repair to bridge with an open grid steel deck and new floorbeams to replace the timber deck.

1969   Designated an American Society of Civil Engineers Landmark.

1975   Designated a National Landmark by the National Park Service.

1983   Rehabilitation featured a rewrapping of the main cables with a neoprene wrap which has proved to be unsatisfactory.

1999   Celebrations to mark the 150th Anniversary of the Wheeling Suspension Bridge.
Completion of a multi-million dollar preservation project.

Selected Bibliography

Bender, Charles. "Historical Sketch of the Successive Improvements in Suspension Bridges to the Present Time," *Transactions, American Society of Civil Engineers* 1 (1877).

Burt, Olive W. *The National Road: How America's Version of a Transcontinental Highway Grew Through Three Centuries to Become a Reality.* New York: John Day Company, 1968.

Cordier, J. *History of Interior Navigation (Histoire de la Navigation Interieure).* Paris: Fermin Didot, 1820.

Davis, Gene D. *Charles Ellet, Jr.: The Engineer as Individualist, 1810-1862.* Urbana, IL: University of Illinois Press, 1968.

Douglas, Howard. *Military Bridges.* London: John Murray, 1853.

Drewry, Charles Stewart. *A Memoir on Suspension Bridges.* London: Longmans, Rees, 1832.

Ellet, Charles, Jr. "Bridge Across the Potomac." *U.S. History* 23:3 (April 1972), 221.

_____. *Report on a Suspension Bridge Across the Potomac for Rail and Common Travel.* Philadelphia, John Clark, 1852.

_____. *Report of the Wheeling and Belmont Suspension Bridge.* Philadelphia: John Clark, 1847.

_____. *Report on the Mississippi Bridge.* Philadelphia: William Slavely and Co., 1840.

_____. "Suspension Bridges—Plan. Plan of the Wire Suspension Bridge About to be Constructed Across the Schuylkill at Philadelphia." *American Railroad Journal* 4:5 (March 1, 1840, New series).

Finley, James. "A Description of the Patent Chain Bridge." *The Port Folio* 3:6 (1810), 441, also published in Uniontown, PA (1811), William Campbell.

Hopkins, H.J. *A Span of Bridges.* Devon, England: David and Charles, 1970.

Juan, George and Antonio DeUllos. *Voyage Historique de l'Amerique Meridionale.* Spanish, 1748; translated to English, London, 1758.

Kemp, Emory L. "Ellet's contribution to the Develop of Suspension Bridges." *Engineering Issues—Journal of Professional Activities* 99, New York, ASCE (July 1973), Paper 9872, 331.

_____. "Samuel Brown: Britain's Pioneer Suspension Bridge Builder." *History of Technology* (London), 1977.

Lewis, Clifford M. "The Wheeling Suspension Bridge." *West Virginia History* 23:3 (April 1972), 221.

McCullough, David. *The Great Bridge.* New York: Simon and Schuster, 1972.

Monroe, Elizabeth Brand. *The Wheeling Bridge Case: Its Significance in American Law and Technology.* Boston: Northeastern University Press, 1992.

Morse, Joseph E. and R. Duff Green. *Thomas B. Searight's The Old Pike: An Illustrated Narrative of the National Road.* Orange, VA: Green Tree Press, 1971.

Navier, Claude L.M.H. *Memoire sur les Ponts Suspendus—Rapport à Monsieur Becquey.* Paris, 1823. Later published by Carilian-Goeury, Paris, 1830.

Needham, J., *et al. Science and Civilization in China.* Vol. 4, Cambridge University Press, Cambridge, England, 1971.

Pope, Thomas. *A Treatise on Bridge Architecture.* New York: A. Niven, 1811.

Sayenga, Donald. *Ellet and Roebling.* York, PA: American Canal and Transportation Center, 1983.

Steinman, D.B. *The Builders of the Bridge.* New York: Harcourt, Brace and Co., 1945.

Telford, Thomas. *Life of Thomas Telford Written by Himself.* London: Payne and Foss, 1838.

U.S. Congress. House. *Bridge Across the Potomac, at Washington.* 23rd Congress, 1st session, 1834.

Vicat, M. "Ponts Suspendus en fil de fer sur le Rhône, Rapport." *Annales des Ponts et Chaussées* 1 (1831).

≋ **About the Authors** ≋

Beverly Balch Fluty is a Connecticut native who has always enjoyed local history. When the family was living in Denver, Colorado, Beverly joined the volunteers of the Colorado State Historical Society. Her largest project was researching all the buildings in a run down area of the city which was just being transformed into an upscale development named Larimer Square. She wrote a walking tour, organized volunteers and was chairman of the project until the Flutys were transferred to Wheeling in 1968. Beverly immediately became involved in local heritage projects and spent many years working on the restoration of the Wheeling Custom House, now named West Virginia Independence Hall, which was the birthplace of the State of West Virginia during the Civil War. She has researched the city's history and has worked on publications, exhibits, tours, special events, etc.

Beverly has always been an ardent admirer of the Wheeling Suspension Bridge and coordinated the successful campaign of the Friends of Wheeling, the local preservation organization, to save the Suspension Bridge.

Mrs. Fluty was an advisor to the National Trust for Historic Preservation and vice-chairman of the advisory committee of the Mid-Atlantic Region of the National Park Service and serves on local heritage organizations. She has received national, state and local awards for her work.

Emory L. Kemp is the director of the Institute for the History of Technology and Industrial Archaeology, at West Virginia University. His formal education at the University of Illinois and the Imperial College of Science and Technology in London is in engineering. Previous to his present position he served as chairman of the Department of Civil Engineering, West Virginia University, where he was active in teaching and research in structural engineering. Before coming to West Virginia University in 1962, he was an engineer with leading consulting firms in London.

Professor Kemp's interest in the history of technology, industrial archeology, and the historic preservation of engineering works stretches over more than three decades. He is a founding member of the Society for Industrial Archeology and served as the first editor of the journal, *IA*. He is also a past president of the Public Works Historical Society.

He has organized, led and served as consultant for industrial archeology projects around the country, including those of the Historic American Engineering Record. His special interest is the history of engineering during the Industrial Revolution with emphasis on the history of structural engineering. He was a fellow of the American Council of Learned Societies to study the Industrial Revolution in Britain, 1975-76, and was a Regents' Fellow at the Smithsonian Institution during 1984-85. He has been elected a Fellow at the University of Edinburgh, Scotland, for 2000.

The bridge was still susceptible to high winds as described by a newspaper article on Christmas Day, 1871. The bridge "reeled" during strong winds so that "pedestrians and vehicles were afraid to venture out upon it." This problem was remedied in the 1870s by the addition of stay cables. The suspenders are hung vertically from the main cables while the stay cables are at an oblique angle which effectively strengthen and stiffen the bridge. This photograph shows the system still in use over 100 years later. GERALD S. RATCLIFF

A dramatic view of the Wheeling Suspension Bridge showing the marked contrast between the massive stone towers and the delicate cables. WILLIAM E. BARRETT

An unusual view of the tower saddles supporting the cables. WILLIAM E. BARRETT

On June 20, 1863, the grand finale for the celebration of West Virginia's statehood was fireworks from the Wheeling Suspension Bridge. The Bridge is often used as a spectacular backdrop for fireworks as shown on this post card of the United States Bicentennial celebration on July 4, 1976. The events on November 7, 1999, celebrating the 150th anniversary of the Wheeling Suspension Bridge, featured fireworks, a laser show and the debut of the decorative bridge lighting. The breathtaking lighting display will be a lasting reminder of the special anniversary. PHOTO BY ROBERT WILLIAM BELL, ELLEN DUNABLE COLLECTION